EGGS, EARTH AND SPACE

CLIMATE CHANGE AND THE HEALTH OF OUR PLANET

VITTORIO DE COSMO

TRANSLATED FROM THE ITALIAN BY MARINO D'ORAZIO

Published in 2024 by
Saratoga Springs Publishing, LLC
Saratoga Springs, NY 12866
www.SaratogaSpringsPublishing.com
Printed in the United States of America

ISBN-13: 978-1-955568-38-8
ISBN-10: 1-955568-38-3
Library of Congress Control Number: 2024905717
Translation Copyright © 2024 by Marino D'Orazio
Original Italian Title: Uova, Terra e Spazio
Copyright © 2021 by Vittorio De Cosmo

Translated from the Italian by Marino D'Orazio
Written by Vittorio De Cosmo
Graphic Design by Emily Brooks
Publisher & Book design by Vicki Addesso Dodd

Saratoga Springs Publishing's books are
available at a discount when purchased in quantity for
promotions, fundraising and educational use. For additional
information, book sales or events, contact us at
www.SaratogaSpringsPublishing.com or
mdorazio@doraziolaw.com

"To my children and grandchildren."

INDEX

Soil for life

The Cycles

The Biosphere

Aristophanes and Global Warming

Climate Change

Conclusion

Notes

Author's Note to the Second Edition

This second edition of Eggs, Earth and Space seeks to update the text with the most recent scientific and "political" findings. Unfortunately, these new findings do not improve the overall picture described in the 2008 edition, and in fact worsen it. Essentially CO_2 levels in the atmosphere have increased more than what previous models described, while the various international conferences (COP's) have ended with optimistic group photos and well-intentioned press releases, but with few agreements. In the political arena new actors have taken center stage and announced refusals regarding lowering greenhouse gas emissions – the USA among them. More coal mines have been opened and there has been further cutting of the Amazon rainforest, not only in Brazil but in other "amazon" countries. To further remind us of our fragile existence on our beautiful planet an unexpected global pandemic has taken seven million lives to date.

Is Nature trying to make us understand that we're doings things we should not be doing?

Finally, I take sad note of the passing of the author of the preface to the first edition of this book, professor Giovanni Bignami, the president of the Italian Space Agency for many years, a true man of science, whom I greatly esteemed.

PREFACE

The Intergovernmental Panel on Climate Change (IPCC), a body of the United Nations has concluded that the bulk of available scientific data demonstrates that the warming of the Earth-system is "unequivocal": beyond doubt.

The rise in the median temperature of the atmosphere and the oceans, the melting of the glaciers, increased sea levels, are all scientific certainties. And it is also scientifically certain that the cause of these effects is human activity.

This information, without alarmism, but with scientific rigor, must be made available to everyone, young and old, but especially to the young. A serious scientific exposition in this field is therefore essential.

Vittorio de Cosmo's Eggs, Earth and Space, is a book aimed specifically at young people, whose purpose is to expand their knowledge of the great themes of Earth Science, using Space, and in particular the International Space Station, as a point of reference.

Space is an advantageous place from which to study and understand our planet. And it is just for this reason that the Italian Space Agency (ASI) has always had as its principal aim the improvement of our understanding of the system Earth.

To this end ASI is realizing or about to realize, independently of its participation in missions of the European Space Agency (ESA), a number of activities: the launch of the first two COSMO/SkyMed satellites (with two more to be added in the next few months), which will permit us, with sophisticated radar, to open a window on our understanding of natural emergencies; the upcoming ROSA mission, which will put into orbit a sophisticated instrument installed on board various international missions, which by analyzing signals from GPS satellites, will allow us to study with great accuracy the physical characteristics of the atmosphere everywhere on Earth, thus giving us a better understanding of Climate Changes; the future PRISMA space mission which will enable us to study, in detail never reached until now, the physical-chemical characteristics of the Earth's surface and its vegetation.

A captivating and lively story line, with fragments of mythology mixed in, and two wild and nosy teenagers as protagonists, provides the central thread for Dr. De Cosmo's rigorous scientific work. The author takes his young reader by the hand and guides her with wit and humor along a scientific journey ranging from the Earth's energy balance to el Niño, from thermodynamic machines to climate change.

This method of imparting scientific knowledge has proved itself enjoyable, yet no less serious, not only for me, but has also impressed the jury of the national literary prize, La Citta dei Sassi, which in 2007 awarded it its prize in the essay category.

Eggs, Earth and Space is one of a few examples of popular scientific literature in the field of Earth Science. Yet this is a field which I believe deserves much more attention, because what we know about the physical system of the Earth foreshadows a very disquieting future.

The near future belongs to today's young. That is why educating them and arousing their curiosity will not only stimulate them scientifically but will provide them with the understanding needed to become the adults of tomorrow.

Rome: May 21, 2008
Giovanni Fabrizio Bignami
President of the Italian Space Agency
Member of the Accademia dei Lincei

1

THE BEGINNING

1.1 - The eggs idea

My name is Leonardo, Leo to everyone, and I'm about to tell you the most incredible story you've ever heard. It's even more incredible because, except for some minor details, it's all true.

When all these things happened, I was a student in a Science High School. My class had participated in a competition co-sponsored by the European Community and the European Space Agency (ESA) to come up with an idea for a biological experiment to be performed on the International Space Station (ISS).

My class won with this idea: a small incubator containing fertilized hens' eggs would be brought aboard and allowed to hatch and give birth to space chicks. Amazing no? The idea was the brainchild of my friend Caroline, Caro to us. She's a bit of a screamer, and a real troublemaker, but I like her a lot.

When Caro first fired off her idea, my friends and I practically collapsed on the floor laughing. The concept was ingenious but the best part came when our physics teacher told us that since the force of gravity is annulled by centrifugal force on the ISS, everything just floats: all it takes is the slightest movement and you just take off. So think about those poor chicks: at the slightest flapping of their wings they would fly away like eagles - they'd be better off with stumps! And what about their droppings. Where would they wind up? Naturally, everybody came up with crazy images of "chicks in space."

We laughed ourselves silly. But, thanks to our teacher, the plan went ahead. We outlined it on the application we downloaded from the internet and sent it off to the address on the form.

Two months later the school janitor came into our classroom with a big manila envelope bearing the ESA logo. We ran to get our teacher who, as excited as we all were, started to open it very slowly. While opening it she comforted us by saying that even if we got turned down it had nevertheless been a wonderful and enjoyable experience.

She opens the envelope: a mountain of papers. She finds the cover letter and we all watch her start to faint. Of course nobody rushes over to hold her up, instead we all very politely stand aside and make room for her to collapse on the bare floor.

The janitors are all rushing into the room, the principal is screaming, "what have you done? you're a bunch of delinquents." A teacher is whispering that maybe she's pregnant; in other words a real mess.

Finally she recovers, gives an irritated stare at her colleagues, reassures the principal, and blares out: we've won. The "chicks" experiment has been judged to have merit and will be performed. Our prize is a trip to Guyana to witness the launch of the next European space mission.

Hugs, kisses, back slapping, shouts, screams (Caroline), great mutual admiration. The most used expression: We're a bunch of geniuses! Compliments from everyone, including the principal, who says, "I knew right from the start it was a great idea."

So, we all went home, taller than we were when we went to school that morning by at least a couple of inches, naturally preceded by the news that we had texted with our "banned" cell phones.

It was great to see our parents so proud. I couldn't understand whether they were more proud of us for having come up with a good idea, or of themselves for having given birth to kids who could actually come up with a good idea.

My mother kept happily kissing me; on the phone my

father after his usual question *"how was school today?"* - which he asked even when there was no school - was speechless for at least half an hour listening to my story. Even he was ecstatic.

At first the days went by very fast, then slower and slower. In school our math teacher, with the help of other teachers, started to prepare us for the trip. Intensive hours of English and French, also Geography and Science.

We were all promoted. Why not? Weren't we all geniuses?

It wasn't easy to find Guyana in our atlas. It's a small country on the Atlantic Ocean near Venezuela, very near the Equator. It started as a prison colony where the French used to send the worst - or who they thought were the worst - criminals. Even the famous Papillon was sent there. But in the movie, both he and his fellow prisoners seemed like good guys to me.

In Guyana, as in other tropical countries, there aren't four seasons, only two: the dry season, when it's hot and never rains, and the wet season, when it's hot and it rains every day. We would get there in July right in the middle of the rainy season. In our geography textbook we learned that the temperature wouldn't be so bad, on average 86°F in the shade.

My grandmother Pina nevertheless suggested I bring something a little heavier, "you never know." I terrorized her on the phone, or so I thought, by telling her that over there the crocodiles peacefully strolled the streets looking for astronauts and students to eat. The itinerary for the trip was amazing.

Day one: Rome to Paris. A bus would be waiting for us at the airport exit and take us to the ESA headquarters which is located in the center of the city.

There we would meet the other student group who also won the competition. They were English. Then, after being welcomed by the Director General of the ESA, we would listen to speech number 1, speech number 2, speech number 3, and then after meeting some European astronauts we would receive some gifts: tee shirts, caps, ESA umbrellas; then finally for whatever was left of the afternoon we would be free to get to know Paris.

Day two: Paris to Kourou, a flight of about 13 hours, check-in at the hotel, and the rest of the day dedicated to rest.

Day three: visit the base, meet the astronauts, visit the Launch Center, return to the hotel.

Day four: Visit the launch pad, look at the awesome Arianne VI rocket and the CRX space capsule, return to the hotel and wait to be picked up again and taken to the base to witness the launch.

Day five: Kourou back to Paris.

Day six: Paris - Rome. What a fantastic program!

The space mission would be very interesting: it was going to be the third flight of the CRX European space capsule. It would ferry three astronauts to the International Space Station where they would be performing some experiments and also mounting some new parts onto the station itself, still under construction. On board the CRX there would also be 5 mannequins as passengers, resembling five normal people (2 adults, 2 teenagers, and one senior citizen). The mannequins were full of sensors which would produce data for understanding if the CRX could transport civilians and not just astronauts.

We spent most of each day thinking about the trip. Many of us had never been on a plane, none of us had ever been in a tropical country. What was a country with only two seasons like? What was Arianne VI like? The space capsule? The astronauts? Would there be a powerful blast at the launch? And what if it exploded in flight? We had so many questions, and even more answers!

One day Caroline and I were alone, talking about all these things, when all of a sudden this volcano screamed out another of her incredible ideas. The big day arrived.

The afternoon before, the President of the Italian Space Agency invited us to a large conference room where, *after his congratulatory speech, he treated us to some good ice-cream, and we were each given a nice Swatch watch. That night nobody slept. Then came the good-byes; our mothers' and everybody else's tears, the airport, the fear of flying, everybody's need to pee, on the bus, in the airport, on the plane before takeoff, during the flight and the landing, and all at the same time; our curiosity about every little thing, our fear of any noise or vibration, all these things were going through our minds at the same time, we were very confused.*

We arrived in Paris. A bus was waiting to take us to the European Space Agency building. There a man greeted us and handed us caps, tee shirts, folders, bags, but no ice-cream.

Finally we got to see Paris: really beautiful!

Then another nervous night, a few more worries about flying for so many hours over the Atlantic with no land below, etc.

We made it! The heat was terrible, we checked in at the hotel. We visited the launch base, and met with the chief in charge of the launch. She was a Spanish woman named Amanda Rivera Fernandez de Fuentes y Avellanas y... A no nonsense very good looking woman. She explained the mission to us.

We met the astronauts: the Commander, Diane Gaiova, unknown nationality, very serious almost military appearance, tiny (maybe they would put a pillow on her seat so she could see out of the cabin), but who inspired lots of trust; the co-pilot: Paul Cadorinne, French, looks like a formula 1 driver, didn't say a word; and a German scientist: Dr. Franz Rushodovisky, glasses, messy hair, disheveled, didn't look anything like an astronaut.

They explained the mission in more detail, their duties on board the CRX and on the ISS, the scientific experiments they were supposed to perform, their daily physical activities, etc.

That morning Caroline had repeated for me the main points of her plan: don't stand in front so they can see you, or in back so you'll be noticed, dress normally, nothing extravagant, mix with the others, in other words blend in with the group.

When we got together with the other English and Italian kids, more than 40, I realized it wasn't too hard to go unnoticed, especially since we were all wearing ESA tee shirts, ESA caps, and all the same jeans and shoes. It was a little depressing to see how we were all the same, it was even hard to tell the difference between guys and girls. And of course we were all wearing sunglasses. We were like a flock of sheep waiting to see the shuttle on the launch ramp. This was going to be the most interesting part of the day.

A very fast elevator took us all up to the hatch of the capsule. They even let us take a peek inside. We were getting on line to come back down when all of a sudden Caroline shoved me, and in a flash I found myself behind the door to the stairs leading down. We locked ourselves inside a closet for about an hour. Then, when we didn't hear any more noises outside, we quietly slipped out. All activity near the shuttle had stopped, maybe because, as my stomach was reminding me, it was lunch time.

We went into the shuttle, we undressed the two "young" mannequins and put them both in the closet we had hidden in. We stripped down to our underwear and put on the light jump suits the mannequins had on under their space suits. We were so scared about what we were doing that we forgot to be embarrassed about seeing each other in our underwear.

We got into the space suits and attached all the electrodes exactly the way they had been attached on the mannequins: two on the feet, two on the legs, one on the stomach, one on the heart, two on the shoulders, and three around the forehead. Before we put on the helmet, Caro said: "so far so good, maybe the surveillance cameras inside and outside the shuttle haven't been turned on yet; now let's put on the helmets and if we don't want to be discovered we have to act just like real mannequins: don't move and don't touch me at all, unless you're about to have one of your infantile panic attacks!"

I closed my helmet and completely ignored her. But I couldn't move, I was scared stiff. I was thinking about what might happen; for sure when the technicians came back they would spot us and kick us all the way down the stairs. Then we'd get a few more from our teacher and when we got back to Italy even bigger ones from our parents. I was doing so much thinking that I finally fell asleep. I have no idea what Caroline was doing all this time, but I bet she was in no better shape than I was. I'm sure she was scared too, and that's why she also fell asleep.

It all happened so fast. A slight vibration, a big noise, and finally a kick, not like the ones I sometimes got in the rear end, but a gigantic earth shattering jolt that pushed my whole body right out of itself: eyes, tongue, heart, stomach, everything! Just before that, my eyes met Caroline's and I whispered, "break a leg." She did the same.

My bloodshot eyes clouded my sight, although to tell you the truth there was really nothing to see. I probably also fainted, then all of a sudden I felt better. I felt something holding my hand, it was Caroline. That devil had sent us on an adventure a lot bigger than us both. I still couldn't figure out how we'd managed to take the place of the two "young" mannequins, and even more unbelievable, how nobody had noticed.

When we finally landed back on Earth they told us what had happened at the base in the meantime. Nobody noticed we were gone! All the students had been divided into different groups, and were involved in so many activities that they never noticed we were missing. So much for my idea that we were so unique that we couldn't go unnoticed.

The only person who suspected something was the launch director Amanda Rivera Fernandez de Fuentes y Avellanas y ...

A few hours before takeoff she asked for checks on all the experiments that were going to be performed. And these included the mannequins and the incubators containing the eggs.

The signals coming from the incubators showed that everything was working right. But the signals from the mannequins indicated a slight malfunction in the two "young" mannequins, a little static in the signals.

The report she got didn't reassure her, and even though she had a lot to do, she couldn't get that static out of her head. She sent a technician to check out the signals. He reluctantly did it in the few free minutes he had before the launch.

What he discovered seemed so weird and so wrong that he thought it best not to say anything to the director unless she asked him first. As the launch time got closer and closer the activity at the site became more and more crazy, but it was organized craziness. Everybody knew exactly what to do and followed all the procedures they had memorized. Sitting at their control stations everybody had their eyes glued to the monitors in front of them while they talked constantly into their headsets, made a lot of gestures, took notes, drank cokes, chewed gum furiously, ate mega sandwiches, etc. etc.

And that's how we got to 5, 4, 3, 2, 1, 0. Fire.

Luckily everything went off exactly as planned.

1.2 - The CRX Space Shuttle

Do you remember the Niña, the Pinta, and the Santa Maria? They were the three ships that carried Columbus and his friends toward the discovery of America instead of the never found Indies. Their principal job was to ferry a bunch of foolhardy sailors from the Earth, as it was then known, to a whole New World yet to be discovered. A world that they couldn't see even from a distance, unlike the way we can now see the distant stars and all the planets.

The technical plans used in the construction of the three ships were simple: They had to be capable of ferrying their crews to an unknown destination hundreds of sailing days away, and to bring them back home; they had to have sturdy very large sails; they had to be big enough to allow the sailors to sail the ships and also have places to sleep; they had to be able to carry enough water and food on board for the voyage out, since supplies for the return voyage would be obtained in the New World, if they ever discovered it.

Modern spaceships are built with similar requirements, in fact sometimes they're even "simpler" since they don't necessarily carry astronauts, only instruments or robots, and they don't have to return to Earth. [Fig. 1.1]

I was a young man when, on a hot July evening in 1969, I followed on black and white television the landing on the Moon of the APOLLO 11 shuttle that carried the three astronauts, Armstrong, Collins and Aldrin. Seeing the footprints of man's first steps on the Moon was an unforgettable moment. [Fig. 1.2]

Looking back with today's eyes at the Apollo spaceship or one of the first Russian Soyuz capsules, one can't help wondering how men were able to survive in such constricted space for such a long time: those were truly heroic undertakings.

Now things are a lot better. The available space inside a spaceship is able to provide a comfortable environment for the astronauts; they don't even need a space suit once in orbit, and can just wear shorts and a tee shirt. The only remaining problem that's not going to be easy to resolve is weightlessness while in orbit. In fact, as we'll see later, on board orbiting spaceships the force of gravity that pulls objects down is offset by the centrifugal force that pushes them outward; and since there's no up or down inside a space ship, objects can't fall and are able to freely float around.

Think about the complications! A glass can't hold water. You can't stop unless you find something to grab. And trying to urinate like you're used to, very embarrassing!

Figure 1.1: *Source, NASA. The CASSINI probe during the integration phase. The CASSINI mission is a combined NASA, ESA, and ASI mission for the exploration of Jupiter and Saturn.*

As of 2022 the only space shuttles capable of bringing human crews into space are the modern SpaceX Crew Dragon and the old small, yet very reliable, Russian Soyuz space craft. The Shuttle STS (Space Transportation System) missions have been the major contributor to the construction of the International Space Station.

The Shuttle was a space capsule capable of carrying a maximum crew of seven people and many tons of material into orbit around the Earth. Its dimensions were such that it could carry a bus. One of its principal advantages was that it could be used multiple times. The first Shuttle, named Columbia, had its first flight in 1981.

Figure 1.2: *Source, NASA. Man's first steps on the Moon (July 20,1969).*

While orbiting the Earth the Shuttle was able to undertake different missions. It could place satellites into orbit or reel them in to make repairs. It also functioned as a real scientific laboratory, and some of them were even able to "dock" at the International Space Station to drop off or take home people and machines. The Shuttle's return to Earth, although much more dangerous, occurred much like the landing of an airplane. [Figs.1.4 and 1.5]

The Shuttle was launched by having it strapped onto an enormous fuel tank the height of a fifteen story building. [Fig. 1.3] The tank with the Shuttle on its back was carried to a height of about 50 kilometers (31 miles) by two rockets, called boosters, attached to the sides of the tank. [Fig. 1.6] When the boosters exhausted their thrust, they detached and fell into the ocean (where they would be recovered for future use). At this point the three Shuttle thrusters, fueled by the liquid Hydrogen and Oxygen stored in the giant fuel tank, ignited. It took about 10 minutes for the Shuttle to achieve orbit, and at that point the fuel tank was detached and abandoned to its own destiny. Now the Shuttle was free to begin its mission. Like one of Columbus's ships, the Shuttle was divided essentially into three parts: the command zone, the living quarters, and the large cargo bay where the scientific instruments or satellites to be placed in orbit were stored. It's important to remember that the Hubble Space Telescope, which has made it possible to discover and understand so much about our Universe, was placed in orbit and then repaired by Shuttle mission STS-125.

At times the road to the conquest of space has been marked by tragic events, and in fact a number of astronauts have lost their lives in this enterprise. The launch and re-entry phases posed particular danger. On launch the danger was that the rocket carrying the shuttle might explode. Re-entry was also extremely dangerous because friction between the shuttle and the earth's atmosphere caused intense heating of the shuttle's outer shell. For this reason the shell was coated with ceramic tiles able to withstand temperatures in excess of 1000°C (1800°F). If, due to a maneuvering error or other malfunction, the temperature were to exceed these limits, the shuttle could even burn up.[1]

1.3 - The International Space Station (ISS)

The space flight of the Russian Yuri Gagarin in 1961 - the first man to orbit the Earth - started the era of human conquest of Space, which culminated with men landing on the Moon.

The great economic sacrifices made in order to land on the Moon led to the realization that in order to continue in the exploration of space the world needed to find new strategies. The first step was the construction of a base in orbit around the Earth financed with the contribution of the world's richest nations - for the benefit of all humanity: the International Space Station (ISS).

Figure 1.3: *Source, NASA. The Space Shuttle in its launch configuration.*

The precursor of the ISS was the Russian Space Station MIR. The International Space Station is the largest construction project man has undertaken in Space. Participating in its construction are the American Space Agency NASA, the Russian Space Agency, the European Space Agency ESA, the Canadian Space Agency CSA, the Italian Space Agency ASI, and many others, totaling 16

Figure 1.4: *Source, NASA. March 9, 2009. The Shuttle is being readied for launch at the Kennedy Space Center in Florida for mission STS-125 toward the Hubble space telescope.*

Figure 1.5: *Source, NASA. The Space Shuttle Atlantis, mission STS-117, landing at Edwards Air Force Base (USA) following a successful fourteen-day mission to the International Space Station in 2007.*

Figure 1.6: *Source, NASA. Launch of Space Shuttle Atlantis, mission STS-81, in 1997.*

different countries.

Its assembly in Space, begun in 1998, occurred in various stages. It was built with modules constructed on Earth and ferried into orbit by the Space Shuttle or the Russian Soyuz spacecraft. The assembly of the modules in space took many years, and occurred in different phases, beginning with the living and command quarters and followed by those needed for scientific work and experiments. The dimensions of the ISS are gigantic. It is approximately

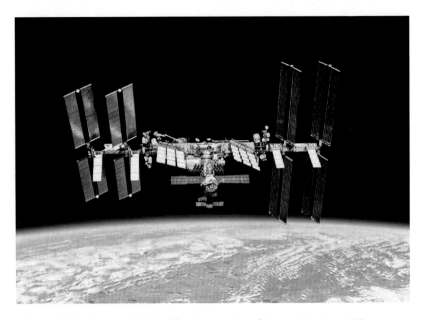

Figure 1.7: *Source, NASA. The International Space Station, ISS.*

110 meters long (120 yards), 70 meters wide (76.5 yards), and weighs approximately 500,000 kilos (1,102,000 lb). The surface of the solar panels, which provide the necessary electricity to power the station, measures by itself half of a football field. The station is so big that on a clear night its general structure can be seen with binoculars. [Fig.1.7][2]

The ISS can host seven people, but in order to function it requires at least three on board. It moves around the Earth in an almost circular orbit varying in height between 250 and 450 km (150 - 280 mi).

The main purpose of the ISS is to conduct experiments which cannot be performed on Earth. In its six laboratories experiments in all fields of science are performed, ranging from astronomical observations to research on osteoporosis. In addition, all the experiments on the effects on the human body of long term exposure to zero gravity are also performed. This research is fundamental to the establishment of a human colony on the Moon or a manned trip to Mars.

The ISS was to be completed by 2004, but because costs have exceeded estimates, and because it hasn't been utilized as much as anticipated (due in part to the Shuttle tragedies), it was only completed in 2017. It is scheduled to be operational until 2024, and to be dismantled by 2028.

2

WHY DO WE NEED TO STUDY THE EARTH?

2.1 - Aboard the ISS

Tradition dictated that once the capsule achieved orbit, back on Earth the launch director would light a gigantic Havana cigar as a sign of her satisfaction.

But just as Amanda Rivera Fernandez de…was about to light her cigar she remembered to ask the technician if he had analyzed the noise coming from the two mannequins.

They told me that all hell broke loose at that point. The technician, somewhat red in the face, said that he hadn't had a chance to spend too much time on the problem, but he was sure that the electrical noise was probably due to some minor malfunction of the sensors, surely nothing too important. He said that he had analyzed those strange signals with his "signal analyzer" and had discovered that - at this point he started laughing, making what he was about to say even more incredible - they resembled the heart beats of a person sleeping. And he was still laughing when the chief let out a high pitched ferocious homicidal scream that reduced him to a little mound of jelly quivering on the ground.

After her initial explosion, the director got control of herself and with the sense of absolute authority she came out of the womb with, ordered the technician to go and count the number of kids present at the launch, and to do it precisely, on pain of a cruel premature death. She then asked another technician to analyze the signals coming from the five mannequins and report back immediately.

The answers came back after only a few minutes. The count showed that there were two Italians kids less than the original number present at the launch pad. The technician, now in agony, assured her that he had personally counted them at least three times.

The Italian teacher, desperate and in tears, named the two missing students: a certain Caroline and a certain Leonardo.

The other technician also had some surprising results. The electrodes placed near the heart of the five mannequins had produced different signals during the launch. While those coming from three of the mannequins, after initial strong oscillations during launch, had gradually diminished in amplitude, finally stabilizing to minor oscillations, the signals coming from the other two mannequins had behaved like those coming from the astronauts.

The chief at this point understood everything. She threw her unlit cigar hitting the poor technician and yelled: "Damn it...We have a couple of stowaways in the capsule." Damn it! Damn it! Damn it!... and so on louder and louder

Figure 2.1: *Source, NASA. The Space Shuttle Discovery docking with the ISS (June 9, 2002).*

for a couple of minutes. Finally after the last "Damn it" she said: "We're lucky. They could have been a couple of terrorists!

Figure 2.2: *Source, NASA. Photo taken on February 24, 2005 from the ISS at an altitude of about 450 km (c.280 miles) above the Southern Pacific Ocean.*

When we bring those delinquents back to Earth I'll flatten their behinds for them."

It was decided to inform the commander Gaiova only after the spacecraft had completed a few "peaceful" orbits before docking with the Space Station.

The docking of the CRX with the ISS was stress free and went off perfectly, as was the transfer of those on board to the Station. We were greeted by five other astronauts, all from different countries, who were very glad to see us because it meant that in a few days some of them would be going back to Earth. They had some fun looking us over but weren't too interested otherwise.

Living in the Space Station isn't easy. In the first few hours you get these terrible headaches, the temperature isn't pleasant, and let's not talk about those awful smells. But after a while we got used to it and started floating around doing all sorts of weird cartwheels and generally having a great time. Weightlessness was great. We were floating in space like balloons. All we had to do was will ourselves to move and we just took off.

Figure 2.3: *Source, NASA-GSF. Eruption of Mount Etna (Sicily). Photo taken from the ISS on August 6, 2006.*

But we missed our mothers! They listened to all our miseries, and after they screamed at us they always gave us bread and Nutella. Our mothers always forgave and forgot.

Instead up there, after ignoring us at first, everybody was mad at us. The toughest one, the most cold blooded, who could cut like a razor, and who had the ability to make us feel like crap was Gaiova, the commander. I'm sure she'd been to military school. She said very little to us, she looked at us even less, but just that little bit made us feel sick.

After we got completely reamed out by everybody down on Earth and especially in space, the atmosphere finally seemed to calm down when the commander Gaiova pawned us off onto Doctor Franz Rushodovisky.

On orders from the commander Gaiova they sealed us off in one area of the Space Station like babies in a sand box. Not only couldn't we follow the activities of the astronauts but we couldn't get anywhere near them. I guess they were right.

After doing a lot of weightless somersaults we passed the time looking down on Earth from whatever port holes we could look out of. The view took your breath away. A beautifully colored ball in the middle of a black sky full of stars. And the view of the part of the Earth where it was night was even better: it was all black and studded with a million tiny points of light.

Since the ISS orbits the Earth every ninety minutes or so, we passed over the parts where it was day and the parts where it was night every forty five minutes. We could see the clouds, the oceans, the land, the lights at night, it was amazing.

Somewhere down there were our parents, more scared than we were, our friends who envied us (maybe), and everybody else. We could see them all at once, rich and poor, good and bad, black and white; who was sleeping, who was waking up, who was eating, who was working, everybody. You couldn't make them out but they were all there. Everyone except us of course, we were on board the ISS.

The Doctor didn't want to have anything to do with us. He just ignored us. He had bigger and better things to do and didn't want us under foot. Caro thought we should stay on his good side, or at least let him know that we existed; so she started asking him little stupid annoying questions that at first really irritated him, but then gave him an excuse for putting us to work.

Especially annoying were her questions about the chicks. When she said, "where's the incubator with the eggs? Did you know it was my idea?" the Doctor's face turned beet red and blew up like a balloon. Then, because he was making a bunch of hand gestures like he was ready to hit somebody, he couldn't stop himself twirling around in the air like a dog chasing its tail.

Caro and I grabbed on to some supports and hurried over to grab him. After we steadied him, she shamelessly asked him how he could possibly lose control like that, and that he should make a better effort to coordinate his movements if he

36

didn't want to get hurt or get yelled at by the "general," our nickname for the Commander.

2.2 Why study the Earth from Space?

The Earth is a unique planet in the Solar System and undoubtedly planets like the Earth are very rare in the Universe. The singularity of the Earth lies in the fact that within a layer approximately 10 km above and 10 km below sea level a certain phenomenon takes place, which as far as we know is unique: Life happens.

This layer, about 20 km thick, is called the Biosphere.

The Sun is the only source of energy which feeds our Planet. The Earth's distance from the Sun allows the energy it receives to maintain equilibrium among Soil, Water, in its three phases: liquid (fresh and salt water), solid (ice), and gaseous (water vapor), and Air, thus creating the most suitable conditions for life to grow and thrive.

The obvious question then arises: if the Earth is such a perfect system why should we worry about it? Why study it with so much interest?

The answer to these questions is simple: wonderful natural things are happening in the Biosphere, but also terrible ones (storms, hurricanes, cold, heat, volcanic eruptions, earthquakes,tsunamis, etc.); and we want to be able to forecast these events so that we can prevent or mitigate their effects.

Add to this the strong conviction in the scientific community that human activity is causing damage to this delicate equilibrium among soil, water and air. This damage may cause short term changes in the Biosphere, as we are witnessing, and long term changes which will be experienced by our children and grandchildren - changes that will make life on Earth as we know it very different, if not impossible.

The study of the Earth as system is therefore very important, but not easy by any means.

It is relatively "easy" to study the evolution in time and space of a physical system composed of a few particles. Take for instance a satellite in orbit around the Earth; we can predict where it will be in an hour or a year with great precision.

Air and Water on the other hand are fluids, physical systems composed of an enormous number of particles. In such a system it would be impossible to know what happens to a single particle. But it is possible to understand the average properties of the component particles as a whole; therefore we can only predict their behavior in global form. What this means for example is that if we are studying a cloud, we can estimate how many droplets are in it, and at what average speed they are moving, but we cannot know what the speed or trajectory of a single droplet within the cloud will be in the next two minutes. To these difficulties, typical of all fluids, we can add the following problems related to terrestrial fluids:

Temporal: Air and Water are subjected to very rapid phenomena, like variations in solar energy, magnetic storms, etc., and sometimes to periodic and slow changes: the day, the lunar month, the seasons, the movement of the Earth's perihelion, etc. In these last examples changes occur over long periods of time in relation to the average human life span, so that their effects are verifiable only by future generations.

Spatial: The spatial dimensions over which these phenomena occur are extremely large, since they involve a great portion or all of the Biosphere.

Uniqueness: The difficulty of studying these phenomena is made greater by the fact that each one is unique. Let's suppose, for instance, that we have developed a theory about a certain event (for example, a storm over Rome on September 28th at three o'clock in the morning). Since there won't be another event exactly the same, we can only verify our theory using events that are similar but that occur in other places, or at other times of day, or in other periods, or all of the above.

Until about fifty years ago it was very hard to study phenomena occurring on Earth since scientists were confined to Earth, to being inside the Biosphere. Being inside a phenomenon permits us to have an exact understanding of it, but does not permit us to understand it in its completeness or in global terms. In order to do so we must be outside of it. This is why it's so important to study the Earth not only by being on it, but also by being outside of it, in Space: seen from a distance we can study the Earth with a global vision. Since Space has become accessible it has proved to be a privileged place for the observation and study of the Earth System. Being able to globally observe the dynamics that govern our planet has opened new frontiers for its understanding. Parallel with the conquest of Space, considerable technological and scientific research has been carried out for the development and manufacture of instruments, intended to be loaded onto satellites, capable of providing scientifically accurate data in order for humanity to tackle the enormous challenges it will face and need to manage.

Once we establish which phenomenon we want to study, our task then is to figure out the best orbit, which instrument to use, how to transmit and receive the data produced by the instrument, how to verify whether our theoretical model of the phenomenon is or is not correct.

2.3 - Some notions about Electromagnetic Waves

Before continuing it may be a good idea to provide a few ideas about the properties of electromagnetic radiation.

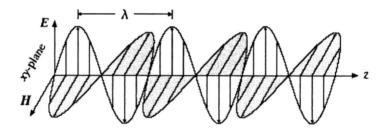

Figure 2.4: *Diagram of an Electromagnetic Wave. The Electric field is always perpendicular to the Magnetic field, and both are perpendicular to the speed of propagation, which in a vacuum is the same speed as the speed of light.*

All electrically charged particles, for example electrons, when they are subjected to rapid acceleration, as when they oscillate, produce energy in the form of electromagnetic radiation. This energy is transmitted via so called Electromagnetic Waves. These waves differ from other types of waves, such as sound waves, because they do not need a medium to spread; they are oscillations of the Electric field and the Magnetic field.

In empty space, electromagnetic radiation travels at the speed of light [c], about 300,000 km/s (186,000 mi/s); when it moves in other mediums its speed decreases. The relationship between its speed in a medium and in a vacuum is the refraction index of the radiation in the medium.

As with all wavelike movements, electromagnetic radiation is characterized by its wavelength [λ] (the distance between two successive crests of the wave), and its frequency [υ] (the number of times in which the electromagnetic field oscillates), and it is measured in Hz. [Fig. 2.4] These two characteristics of electromagnetic radiation in a vacuum are tied by a simple equation:

$$\lambda \upsilon = c$$

In other words the speed of an electromagnetic wave is the product of its length and its frequency.

When a band of electromagnetic waves moves, it carries energy which is dependent on the frequency and intensity of the band. In order to better understand this phenomenon it helps to think of a band of electromagnetic waves as being made up of packets of energy called photons, each with energy:

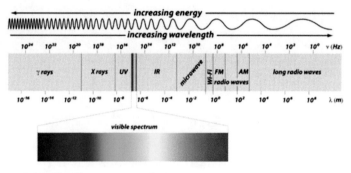

Figure 2.5: *The Electromagnetic Spectrum.*

$$E = h\upsilon$$

Where υ is the frequency and h is a constant universally known as the Plank constant.[3]

The higher the frequency the more energized the photon. This model gives us a much more intuitive description of electromagnetic radiation. For example the intensity of a band of electromagnetic radiation hitting a given surface for a certain period of time is the same as the number of photons hitting that same surface for the same period. [Fig.2.5]

One way to make electrons around the nucleus of an atom oscillate, thereby causing them to emit electromagnetic radiation, is to provide them with energy by heating them.

In nature all bodies with a temperature above Absolute Zero[4] (-273.15°C or - 459.67°F) emit electromagnetic energy at all wavelengths. Depending on the

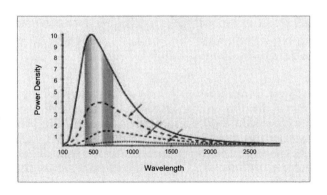

Figure 2.6: *Spectral distribution of the electromagnetic energy emitted by a Black Body at different temperatures.*

characteristics of the body and on its temperature, the distribution of energy along various wavelengths, also known as spectral distribution, will vary, but will never be greater than that of so called Black Bodies, that is those bodies capable of absorbing all the energy hitting them.[5] The amount of energy a black body can emit depends on its absolute temperature [T], specifically the total energy emitted is proportional to T4; in other words a body with an absolute temperature of 600K (about 327°C or 620.6°F) emits 16 times more energy than a body at a temperature of 300K (about 27 °C or 80.6°F).

Figure 2.6 shows the spectral distribution of energy emitted by a black body at various temperatures. As we can see in the figure, as the temperature of the black body increases the wavelength at which the body emits its greatest amount of energy becomes increasingly smaller.

The Sun, which has a surface temperature of about 6000K, has its maximum emissions between wave lengths 0.4-0.7μm[6]; this is in the band of the spectrum called visible light, since we can see it with the naked eye. The Earth, whose temperature is about 300K has its maximum emissions at about 10μm; this radiation is not visible to the naked eye, but perceivable by our bodies as the sensation of hot or cold. This band of the spectrum is called thermal infrared.

Any other type of body will always emit less energy than a black body because it will not absorb all the energy that hits it, but will reflect or transmit a portion. The characteristic manner in which a body reflects and emits electromagnetic radiation is called its spectral signature or body spectrum.

Of all the enormous amount of electromagnetic energy emitted by the Sun only a very small, very select, part is useful for life on Earth. Luckily, this job of selection is done by the atmosphere surrounding the Earth. The atmosphere acts like a filter, reflecting the electromagnetic radiation at certain wavelengths, transmitting others, and attenuating still more. This atmospheric behavior, as we'll see later, is

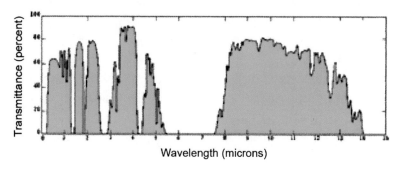

Figure 2.7: *Atmospheric transparency at different wavelengths. Atmospheric windows are in blue.*

due to the absorption and diffusion processes of the molecules and particles which make up the various components of the atmosphere. The portions of the spectrum where the atmosphere is transparent are called "spectral windows."

As we can see in Figure 2.7, the atmosphere is transparent in the visible range (0.4-0.7 μm), in the mid infrared range (3-5 μm), and in the thermal infrared (8-12 μm), and in the millimetric and radio bands (wavelengths from cm and greater).

Given this selective behavior of the atmosphere any instrument placed in space which aims to observe the Earth's surface must necessarily operate within the bands of the spectrum corresponding to the atmospheric "windows."

2.4 - Studying the Earth from Space: "Remote Sensing"

The Earth is a physical system formed by a number of subsystems which interact with each other in a complex way, such that an event occurring in one subsystem triggers other events in others. In order to monitor these events, we need to understand them and be able to track them in both space and time. Earth observation from space is a formidable instrument for understanding and monitoring our planet in global terms. The principal methodology for this task is called remote sensing, in other words "measuring from a distance."

Remote sensing is the method used to measure the electromagnetic energy emitted or reflected by the Earth.

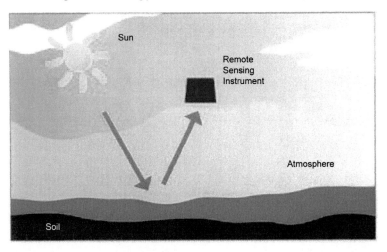

Figure 2.8: *The electromagnetic landscape in which a remote sensing instrument operates.*

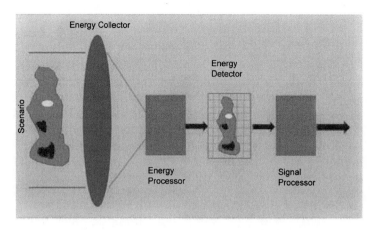

Figure 2.9: *Diagram of a passive remote sensing instrument.*

Figure 2.8 illustrates the typical way an instrument aboard a satellite is used to study the Earth or any other planet. Solar radiation hitting the Earth interacts with the atmosphere and with the area of the Earth's surface under observation, and is reflected, diffused, or absorbed. The radiation which reaches the instrument will measure the effects of these processes as well as the "spectral signature" of the surface being observed. Knowledge of these spectral signatures allows us to identify the different materials which make up the surface area under observation, and to understand their chemical-physical state.

There are two types of Remote Sensing instruments: Active, those which independently illuminate the scene to be studied, such as systems that operate like radar; and Passive, those which use sunlight as illuminator in order to measure radiation reflected and emitted by the Earth.

The "remote sensing" instrument most familiar to us is our eye. The eye, as we've said, is an instrument that can detect only a part of the electromagnetic spectrum: the visible range. In order to "see" in the other bands of the electromagnetic spectrum we need instruments able to detect those specific bands. For example, some animals, like snakes and cats, are able to detect another range of the spectrum not visible to the naked eye, the infrared range.

Regardless of the range of the spectrum in which it operates, a remote sensing instrument is composed of the following parts, as shown in Fig. 2.9:

• an energy collector, such as an antenna or the principal lens of a telescope. There are two kinds of collectors: those that conserve the "form" of what they observe, and those that collect only the energy without conserving the form.

• an energy processor, such as a filter which allows only a specific portion of energy to pass through, or a prism which can separate out its different components depending on their wavelengths.

• an energy detector, which produces an electronic signal in proportion to the amount of energy hitting it. Current detectors are generally composed of a matrix of simple square detectors also known as pixels. Some can even contain several million pixels - think about modern digital cameras which have over ten million pixels. If the collector is capable of saving the form of the observed scene, it can reconstruct its image. Of course the more pixels in the matrix the better the image.

• a signal processor whose task is to amplify, arrange, condense, and record the signals, so as to make them accessible to the operator on earth.

Note that in the case of an active remote sensing instrument we would need to add a block next to the energy processor in Fig. 2.9 for the illuminator. Some of the most important characteristics of a remote sensing instruments are:

• spatial resolution, or the dimensions of the smallest detail we can decipher. Currently we have at our disposal instruments that from space can see details ranging from 1 km, for example weather satellites, to just a few centimeters, for example spy satellites.

• the total field of vision on the earth's surface which the instrument can observe. This dimension varies from a few thousand kilometers, in the case of weather satellites, to a

few kilometers, in the case of high space resolution.
• the spectral range, or the area of the electromagnetic spectrum that the instrument can detect.
• the sensibility, or the smallest amount of energy the instrument can measure.

As previously noted, the solar radiation which reaches a Remote Sensing instrument placed in space has been subjected to two "filtering" actions by the atmosphere: the first incoming solar radiation, and the second reflected radiation reaching the instrument. Of course the atmosphere also "filters" radiation independently emitted by the Earth's surface.

The data produced by instruments aboard satellites is sent by means of transmitters to receiving stations placed on Earth for that purpose. Once the data is processed and stored it is sent to scientists to be changed into easily usable information. [Fig. 2.10]

In simple terms, the data is not transmitted to Earth in a form we use daily, ie. in decimal form, but in binary form, or a series of 1 and 0.

Figure 2.10: *Diagram of data transmission from an instrument on board a satellite to a data receiving station on Earth, and to its users.*

Every digital datum corresponds to the value of the signal as measured by each pixel which constitutes the matrix of the detectors. By reassembling the digital data we obtain the image of the scene observed by the instrument relative to the specific spectral range selected by the block that processes the energy received. Of course the more pixels in the detector the more detailed the image. Modern instruments are capable of deciphering details in images less than 1 meter.

Depending on how they process photons, for example by how many filters and therefore by how many spectral ranges, instruments are divided into Pancromatic, those detecting images only in the visible range, Multispectral, different images of the same scene in various spectral bands, and Hyperspectral, images of the same scene but in numerous contiguous spectral bands.

Sometimes the generation of data by satellites, its transmission to Earth, its processing and transformation

Figure 2.11: *Source, NASA-GSFC. Image of Rome taken from the ISS on the night of April 8, 2015.*

into useful information, is very fast. Try, for example, logging on to www.eumetsat.int and you will see weather images just one half hour after they were taken by the METEOSAT satellite which is in a Geostationary orbit about 36,000 km (22,000 mi) above the Earth.

2.5 - Space missions studying the Earth

Space is a truly advantageous place for the placement of the most advanced instruments used to study the complex phenomena occurring on Earth.

For many years space technologies have been the exclusive dominion of the richest and most militarily committed

Figure 2.12: *Source, EUMESTAT ©2019. On August 26, 2019 powerful storms caused widespread flooding in many areas of Spain. They were quickly identified by the instrument SEVIRI on board the meteorological satellite Meteosat 11.*

countries, that is the United States and the Soviet Union. Fortunately, in time, other countries have also been able to develop space technologies, making it possible to start national space programs or to engage in scientific cooperation with other countries in order to know the Universe better, in particular the Solar System, and to better understand the Earth.

Because of the high costs of space activities and their impact on various sectors of a nation's life, the most advanced countries have Space Agencies, or agencies capable of coordinating and promoting space activities on both a national and international level.

Clearly the best known Space Agency is the American NASA (National Aeronautics and Space Administration); equally important are the Russian Agency called FSA (Federal Space Agency), the ESA (European Space Agency), the French CNES (Centre National d'Etudes Spatiales), the German DRL (German Aerospace Research), the English (British National Space Center), the Japanese JAXA (Japanese Aerospace Exploration Agency), the Canadian CSA (Canadian Space Agency), the Argentine CONAE, the Indian ISRO, the Italian Space Agency ASI, the EUMETSAT (European Organization for the Exploitation of Meteorological Satellites), the American Meteorological Agency NOAA (National Oceanic and Atmospheric Administration), etc.

The primary purpose of all Space Agencies is the study of the Earth as system. With this in mind, Europe has launched the most sophisticated and complete Earth observation system: Copernicus. By visiting the web sites of the various space agencies we can learn about the missions already launched or about to be launched, and about the specific scientific objectives they intend to pursue.

3

ENERGY TO LIVE

3.1 - Who was Empedocles?

After a while the Doctor came back with what we soon found out was his plan to keep us in line.

Do you know who Empedocles was? When we drew a blank, he went on: he was a Sicilian philosopher[7] who about 2500 years ago formulated a theory according to which everything in the world is created by the union and disintegration of four elements: Fire, Air, Water and Earth. Two forces, Love and Hate, are responsible for the "cut and paste" operations among these primordial elements.

Our faces still didn't show the sacred fire that burns in those with knowledge, so, in a thick German accent he started asking us a bunch of silly questions that made us laugh. You really couldn't take this guy too seriously, but we had to play along.

It went something like this:
- You feeling the hunger? You want to eat?
- Yes!
- Why?
- Because our bellies are empty!!!
- Why are your bellies empty?
- Because we've digested everything we ate?
- Why did you digest everything you ate?

- Because we didn't get a lot of food and have been doing weightless somersaults for two hours?
- Ha! So what is the conclusion?
- Don't have a clue!
- Doing somersaults for two hours emptied out your bellies. An empty belly makes you hungry, so you want to eat, so you can do more somersaults.

Here's the translation of his other deductions as we understood them:
- *to make a car move you need gasoline, meaning energy;*
- *to make a spaceship move you need propellant, meaning energy;*
- *to make eggs hatch you need the hen's heat, meaning energy;*

And so on and so forth for a good quarter of an hour.
Let me tell you, when there's no gravity it's dangerous to burst out laughing, like me and Caro did, because if you don't have good control of your movements you'll get blown around like a dry leaf on a windy day. Even the Doctor flew away laughing.
Mother of all conclusions: if you want to do anything you need energy. And what does the Earth need to live? Energy. And what provides the Earth with energy? The Sun.

3.2 - The Energy balance

The Earth is one of the eight planets[8] in our solar system. [Fig. 3.1] It revolves around the Sun, in cold cosmic space (which has a temperature of about -270°C or -454°F), on an almost circular orbit[9] in one year, at a speed of about 30 km/sec. (18.64 mi/sec). The Earth's distance from the Sun is about 149,600,000 km (93,000,000 mi).

This distance is neither too great nor too small, but just right to nourish life on Earth.

The distance between the Earth and the Sun, when compared to the dimensions of our Galaxy (the Milky Way), in which our Solar System is located, is really insignificant. Just think that a ray of light leaving the Sun's surface takes

Figure 3.1: *Source, NASA's Jet Propulsion Laboratory. The Solar system (not to scale). To the right and front of the Sun: Mercury, Venus, Earth and Mars, called the inner planets, and the asteroid belt. To the back: the outer planets Jupiter, Saturn, Uranus, Neptune. Pluto (recently excluded from the planet category) is a comet.*

about 8 minutes to reach the Earth, while one leaving a star in the middle of the Milky Way takes about 35,000 years to reach us.

In order for our planet to "function" it needs energy, and as we've said, this energy is provided by our marvelous star, the Sun, our only energy source. [See Fig. 3.2] The energy we receive from the other stars in our Galaxy is, in comparison, completely insignificant because of their enormous distances from us.

The energy that reaches the Earth is only a minuscule

53

part of all the energy produced by the Sun. About three quarters of the Sun is made up of Hydrogen nuclei (composed of protons), and about one quarter of Helium nuclei

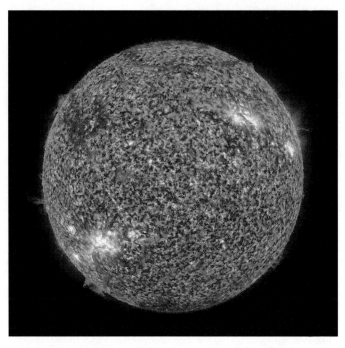

Figure 3.2: *Source, NASA-ESA. The Sun, seen in the Ultraviolet light range through instruments aboard the space ship SOHO on September 23, 2008. In the white circle, a sunspot.*

(2 protons and 2 neutrons). In the Sun's center, because of great pressure, the Hydrogen nuclei are so close together that they are "fused' forming a Helium nucleus. This is the same nuclear reaction that happens during the explosion of an H bomb. From these nuclear reactions, called fusion reactions, in addition to Helium nuclei, an enormous discharge of nuclear particles is produced, accompanied by an equally enormous discharge of electromagnetic radiation. The sun has been producing all this energy for about 5 billion years, and it is believed that it will continue to do so

for a few billion years more (so we can remain calm - in the foreseeable future we won't be left in the dark).

The Sun is continuously radiating this energy into space. The electromagnetic radiation generated by the solar surface encompasses the whole spectrum of electromagnetic waves, with a characteristic spectral distribution[10] that depends on the temperature of the solar surface, which is about 5730°C (10,346°F). At such high temperatures the major portion of the energy is emitted in the so- called Visible range, which is the zone where our eye is most sensitive. In the past, this type of electromagnetic radiation - the only one we knew – was given the name "light."

All the bodies that revolve around the Sun, like the Earth, receive a portion of this energy,[11] whose value depends on their distance from the Sun and on their size. Depending on the surface characteristics of each body a part of this energy is absorbed, while another part is reflected back into space. The absorbed energy causes the temperature of the body to increase. Then, because it now has a higher temperature than the surrounding space, it will radiate energy and thus cool down. The final result is that at a certain point the body stabilizes at a certain temperature; under these conditions the energy absorbed is equal to the energy emitted.

The solar energy deposited in 1 second on a 1m² surface outside of earth's atmosphere and perpendicular to the sun's rays is a full 1368W/m². This value, also known as the Solar Constant, is not really a constant, but varies very little not only due to the variability in a year of the distance between the Earth and the Sun, but also with the sunspot cycle whose period is about 11 years. [See Fig. 3.3]

The tilt of Earth's axis at 23.5°, in relation to the plane of its orbit around the Sun causes solar energy to be distributed uniformly on its surface. This is why we have seasonal cycles and differences in the lengths of days and

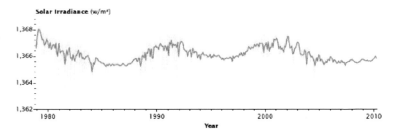

Figure 3.3: *Source, NASA. Variations of the total daily solar constant (bright line) and average monthly (dark line) since 1979. They have not detected a clear trend in the large area but only variations of approximately 1.4 W / m2 due to the cycle of solar heats which is approximately 11 years. (NASA graphic by Robert Simmon, based on data from the ACRIM Scientific Team).*

nights. In fact while the amount of solar energy reaching the equatorial zones is always the same, in temperate or polar zones it varies significantly along Earth's orbit.

In the Northern Hemisphere there is a period (winter) in which the solar energy hitting these zones is less than that hitting the tropical zones, and one in which it is comparable (summer). The exact opposite effect occurs in the Southern Hemisphere.

If we take the average Solar constant over the course of a whole year on the entire Earth, the energy deposited by the Sun in one second on a 1m² surface is 342W/m². This is a very high quantity of energy: if we could, and wanted to, put it to use, it would equal the energy necessary to keep all the light bulbs in a seven-room house lit 24 hours a day (all free!).

On a daily basis, heat, a form of energy, tries to travel from warmer to colder zones with the generous intention of putting everything at the same temperature. This law of nature means that during winter in the Northern Hemisphere warm air and water masses travel from Equatorial zones toward the North Pole. In the summer this movement is greatly reduced since the difference in temperature in the different zones of the Northern Hemisphere is small.

Obviously, the same thing happens in the Southern Hemisphere six months later. This property, typical of fluids - the atmosphere and the oceans are both fluids - allows solar energy to be transferred from warm zones to cold zones, and from those in daylight to those in darkness. Fortunately for us the Earth differs from the Moon, where because there is no atmosphere, the zones exposed to the Sun are extremely hot and those in darkness are extremely cold.

Our atmosphere, located between the Earth's surface and solar radiation, in addition to transferring solar energy from warm zones to cold zones, also acts on the energy reaching the Earth mainly by two other mechanisms: by filtering it and by the so called "Greenhouse Effect."

The atmosphere acts like a giant pair of "sunglasses" on solar radiation, it allows certain wavelengths to pass and blocks the rest, reflecting them back into space. Needless to say, the solar energy that passes through, mainly visible radiation (light) and the energy in the Near Ultraviolet and Infrared ranges, is the kind necessary for life in the biosphere. If this mechanism should fail it would be the end of life as we know it.

A good portion of the energy that passes through is absorbed by the earth's surface (soil and water) while the rest is reflected back into space. The absorbed energy causes the earth's surface to heat and begin to emit energy itself, but at a different, longer wavelength, called infrared radiation. [Fig. 3.4]

As we've seen the amount of solar energy hitting the Earth must be exactly the same as the amount the system reflects and emits; in other words there must be a balance between incoming and outgoing energy. Without this balance we would arrive at the following paradoxical situation: if the Earth took in more energy than it sent out it would, in about 5 billion years, be hotter than the Sun; if the opposite happened it would be colder than anything in existence.

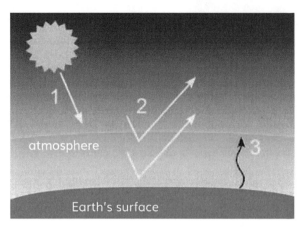

Figure 3.4: *(1) Solar energy hitting the Earth's atmosphere; (2) Solar energy reflected by the atmosphere and by the Earth's surface; (3) Energy emitted by the Earth.*

To better understand the Earth system's energy balance, better known as the "radiation budget," we need to think about a scale with two plates. As long as the weight placed on each side is the same (regardless of the amount) it will always be in balance. In the same way the amount of energy that comes into the Earth system must be the same as the amount that goes out (reflected energy plus emitted energy).[12] [Fig. 3.5]

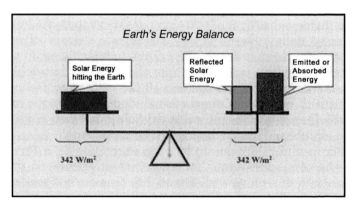

Figure 3.5: *The Earth's Radiation Budget.*

What can change this energy balance? On the left side of the scale (energy coming in), only natural causes, due either to the Earth's revolution around the Sun, the tilt of its axis, or the eleven-year cycle of sun spots.

Variations in the totality of the energy going out (right plate) cannot occur, since the Earth is not a star capable of generating its own energy; but the amount of reflected energy and the amount of emitted energy can change. This is where the Greenhouse Effect comes in.[13]

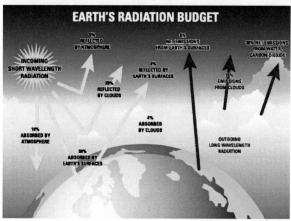

Figure 3.6: *Radiation Budget. The energy balance between incoming and outgoing solar energy. Source, NASA. http://science-edu.larc.nasa.gov*

In addition to filtering solar radiation, the atmosphere is also capable of trapping its heat and not letting it escape.

The lower strata of the atmosphere contain great amounts of water vapor, Carbon Dioxide, Methane, Nitrogen Oxide etc. The molecules in these gases have the capacity to absorb the heat emitted by Earth's surface and to radiate it in all directions. The end result is that the lower strata of the atmosphere, the densest ones, are heated. This heat entrapment is known as the "Greenhouse Effect," and the responsible gases are known as Greenhouse Gases.

Although the amount of energy deposited on Earth by the Sun is considerable, it is estimated that if the Earth had no atmosphere and no Greenhouse gases, its temperature would be like that on the Moon: very hot during the day, and very cold at night. Not conducive to life.

The Greenhouse Effect is therefore indispensable for life, but its increase could break the delicate equilibrium between the amount of solar energy the Earth reflects back into space and the amount it emits.

An increase in Greenhouse Gases, for example an increase in Carbon Dioxide due to greater consumption of fossil fuels (petroleum, coal etc.) by humans, or an increase in biomass combustion (forest fires, agriculture), would trap more heat and cause an increase in the Earth's surface temperature, and consequently bring about global climate changes.

A very important role in the fluctuations of emitted and reflected energy in the Radiation Budget is played by clouds, the best-known component of the atmosphere because the most visible. An understanding of the role of clouds requires detailed knowledge of the manner in which they absorb and reflect short wavelength solar radiation, and how they absorb and release the thermal energy emitted by the Earth.

Low thick clouds reflect received solar energy very well, thus cooling the surface area of the earth below them. High thin clouds, on the other hand, are capable of absorbing the infrared radiation emitted by the earth, thus contributing to local heating.

Thanks to observations made from space we can now affirm that, generally speaking, clouds have the overall effect of increasing the amount of reflected solar energy, thereby contributing to cooling the Earth. It is estimated that if there were no clouds the average temperature of the Earth would be at least 11°C (51.8°F) higher! (Further details in the chapter on the atmosphere).

In addition to clouds, many other natural or man-made factors can cause changes in the reflective or absorption properties of the atmosphere. Among these, of great importance are atmospheric "aerosols", innumerable minuscule particles produced by volcanic eruptions, the burning of fossil fuels, burning biomass, sand, salt etc. Depending on their number, dimension, and chemical-physical properties, these particles can greatly alter the Earth's climate.

Until now we have only considered the role of the Atmosphere in the Radiation Budget, but as we can see in Figure 3.7, we cannot ignore other factors which can cause changes in the reflective characteristics of the Earth's

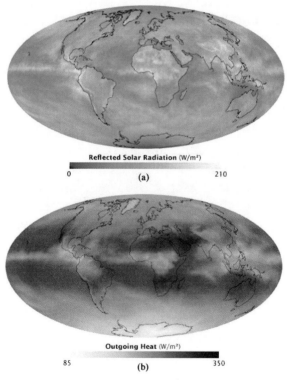

Figure 3.7: *Source, NASA. In the image a) the solar energy reflected from the atmosphere and from the Earth's surface in the Visible and Infrared areas (0.4-5 μm), and b) the energy emitted by the Earth in the Infrared area (5 -10 μm), in September 2004.*

surface; for example changes in snow covered or frozen zones (snow and ice reflect solar radiation almost like mirrors), changes in forested areas, or changes in urban areas, etc.

Since daily weather and long term climate conditions are both controlled by this Energy Balance, one can understand the interest that scientists have been dedicating to this problem for a very long time now. Life on Earth is governed by an extremely delicate equilibrium and modifying it could have dire consequences.

3.3 - Chlorophyll Photosynthesis

What and how do plants eat? How do they breathe? These are questions man has asked himself since ancient times, but a satisfactory answer came only in the last century. Plants are nourished by the solar energy that reaches Earth. Using solar energy, plants are able to convert the Carbon Dioxide and water found in the atmosphere and in the soil into starches and Oxygen. This chemical process, called Photosynthesis, is not only used by plants and algae but also certain ocean microorganisms (Phytoplankton).

Photosynthesis is a chemical process which transforms solar energy into forms of energy used by plants. Plants are the first link in the food chain. They nourish, or provide energy, for other links in the food chain which eventually, in the form of food, reach us, the last link in the chain. As a result, needless to say, we are all indirectly nourished by solar energy.

A fundamental element of photosynthesis is the presence, in certain parts of plants, of pigments like chlorophyll a or b or carotenoids. These pigments are very selective. They absorb electromagnetic solar radiation in the visible range only at well defined wavelengths while reflecting it only at other wavelengths. For example, *chlorophyll a* and *chlorophyll b* absorb blue and red in the violet range, and reflect green, so that leaves appear green to our eyes; carotenoids absorb blue and green and reflect red, so that oranges, carrots and tomatoes look red.

When struck by sunlight these pigments promote a chemical reaction with a dual effect: one allows plants to absorb Carbon Dioxide and release Oxygen and the other nourishes the plant by producing glucose i.e. sugars.[14]

The formula is:

Carbon Dioxide + Water + light $\rightarrow$ Glucose + Oxygen + Water

$$6CO_2 \ + \ 12H_2O \ + \ light \ \rightarrow \ C_6H_{12}O_6 \ + \ 6CO_2 \ + \ 6H_2O$$

Where $C_6H_{12}O_6$ is the chemical formula for glucose.

Photosynthesis occurs in two stages: one, in which sunlight is fundamental, is called luminous, and the other dark. Oxygen is produced in the first stage, Glucose in the second.

Chlorophyll Photosynthesis will be very important if humans are to have the extended stays in space needed to explore and colonize the Solar System. Recent observations indicate that water may be more common than previously thought; in fact it seems that ice may be present on the Moon, Mars, Ganymede and Titan. If these observations are correct it is no longer science fiction to think about creating giant greenhouses on these bodies, where plants, using sunlight and water, will absorb Carbon Dioxide produced by humans and give back Oxygen. Just imagine enormous transparent domes enclosing splendid gardens with fabulous fountains!

3.4 - Darkness: The disappearance of Dinosaurs

Not very long ago, about 65 million years, something happened that radically disrupted life on Earth. At that time the Earth's surface was rich in both vegetable and animal life. It was the era of dinosaurs. These animals, large and small, were the true lords of the planet.

One day a frightening event occurred. An enormous asteroid, a few kilometers in diameter, penetrated the atmosphere and reached the Earth's surface. The impact was awful. In the Yucatan peninsula, where the asteroid hit, we can see the remains of this terrible impact: a crater about 150 km (93 mi) in diameter and a few hundred meters deep. [Fig. 3.8]

The tremendous amount of dust raised by the crash created an immense shield in the atmosphere which solar radiation could not penetrate. In addition many other terrifying global phenomena must have occurred. Some scientists believe that the impact caused great quantities of sulphuric acid to escape, causing extremely violent acid rains; others believe that it triggered a conflagration of global dimensions; and still others that very violent earthquakes occurred in the oceans. Certainly all these events, and especially the absence of solar energy for prolonged periods, had catastrophic consequences for all forms of animal and vegetable life.

Only those species, animal and vegetable, that best adapted to the dark and cold were able to save themselves. It's thought that crocodiles were among the few animals that succeeded in surviving this catastrophe thanks to their ability to remain underwater for long periods of time, in darkness, feeding on the remains of other animals.

Of course there are also other theories about why dinosaurs disappeared. The truth is not yet clear, but one thing is certain: dinosaurs and a many other things living on Earth 65 million years ago disappeared because the environmental conditions which were created (low temperatures, air polluted by dust and gases, absence of light, etc.) were not favorable for the survival of life.

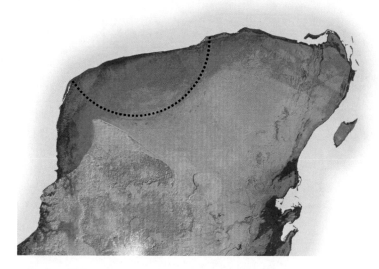

Figure 3.8: *Source, NASA-JPL. The SRTM Mission (Shuttle Radar Topography Mission) February 2000. The SRTM mission was a collaboration among NASA, NIMA (National Imagery and Mapping Agency), DLR (German Space Agency), and ASI (Italian Space Agency). The image shows the Yucatán Peninsula in Mexico. The dotted curve shows a part of the Chicxulub crater formed by the impact of the meteor which many scientists consider responsible for the disappearance of Dinosaurs.*

4

AIR FOR LIVING

4.1 - Why is the sky blue by day and black at night?

"Do you see that transparent stratum of the atmosphere looking blue?" The Doctor was still asking questions. "Well, those are the densest strata." He told us that there was a little bit of atmosphere even where we were, 350 km (280 mi) high: just "a few atoms," but enough to slow down satellites going around the Earth. Because of friction with what little air there is, the Space Station gradually slows down and loses height. When it's at about 250 km (155 mi) they have to give it a boost to get it back to the right altitude.

Here's another of his questions: "Do you know why the sky is blue?" This one took us by surprise. We were so used to seeing a nice clear blue sky that the thought never occurred to us. But when we looked at it from the ISS and saw that it was black as tar, we didn't think the question made any sense.

He told us that the Sun floods all space with energy in the form of electromagnetic radiation and very high energy particles. When the atmosphere gets hit with solar energy it acts like a good pair of sunglasses, it lets some of it through and reflects the rest back into space. Some of the energy that reaches us is the light we can see with our eyes. This light is called "white light", but it's really a mixture of all colors, from blue to red.[15]

One way to understand how light behaves when it passes through the atmosphere is to think about it as being made up of electromagnetic radiation of different wavelengths; the color blue is the light with the shortest wavelength and the color red is the light with the longest wavelength.

The atmosphere is a mixture of different gas molecules and tiny solid particles. When light hits the atmosphere the radiation with the longest wavelengths literally jumps over the gas molecules and continues undisturbed on its way. Instead, the radiation with the shorter wavelengths isn't able to make the jump, so it crashes into the molecules and gets scattered in all directions, in other words, it is diffused. This mainly happens to the wavelengths of the color blue; and that's why the sky is "blue".[16]

Caro wanted to know what happened when light butted into the solid particles. The Doctor answered that since these particles were generally a lot bigger than all the wavelengths in white light, that is of every color, they diffused the light completely. This explains for example why clouds are white. Since clouds are made up of water droplets that are a lot bigger than the molecules in atmospheric gases, they scatter all the wavelengths of visible radiation, meaning all colors, with the end result that they acquire a white light.

After this explanation, and so I wouldn't look like a salami, I made this proud observation: "no wonder the sky is black at night, since the Sun's not out!"

"That's not it my dear friend" said the Doctor. "The answer is a lot more complicated. "Why is the sky black at night? This is a question a young astronomer named Olbers[17] asked himself. It seems that the answer has something to do with the expansion of the Universe. At the beginning of the last century it was discovered that the Universe is expanding, causing galaxies to move away from each other, just the way the dots painted on a balloon do when we blow it up. If the universe weren't expanding, the energy reaching us from all the galaxies would be enormous, and no different day or night. So, the sky at night is black because the universe is expanding."

Caro whispered that this Olbers guy must have been always drunk or stoned out of his mind if all he did was imagine all this stuff and ask himself these questions.

But the Doctor just started talking about the air again: *"Of course even the air has mass. You should know that the mass of a column of air rising from sea level to the end of the atmosphere with a surface area at its base of 1cm² is about 1000 gr (2.2 lbs), and that there are about 600,000 billion billion (6 x 10²³) molecules in about 30 gr (1 oz) of air. With these figures you can have a good time trying to calculate how many molecules there are in the whole atmosphere."*[18]

After the Doctor left, I asked Caro if she had the problem figured out, she said no, but that it was real simple. I had such a confused look on my face that Caro said: *"you're such a numbskull! First you figure out how many molecules there are in a kilo of air in the column, and then you multiply that number by how many centimeters squared the earth's surface is."* I was amazed. She was right. It was simple.

4.2 - The general properties of the Atmosphere

The atmosphere is a fluid made up of a mixture of gases. It encircles the Earth like skin. It doesn't "fly" away because it is kept together by the force of gravity.

A fluid, which can be a gas or a liquid, is a huge combination of molecules (think about tiny marbles) which are in constant chaotic movement due to the effects of their temperature. In addition to this constant disorganized movement of molecules, the fluid as a whole can move in a well defined direction, like water in a river or smoke from a chimney. The "mission" of the molecules in a fluid is to transfer their energy to neighboring molecules, so that in time a thermal equilibrium is reached, and they will all have the same temperature. This energy transfer is made easier by the fact that molecules in a fluid have very little interaction with each other, and therefore, especially in gases, are quite free to move around. An important end result is that fluids do not have a particular form or shape, so that gases will occupy all the volume available, and liquids all the surface available. The main difference between gaseous fluids and liquid fluids is that the former can be compressed while the latter basically cannot.

The atmosphere is the site of many complex phenomena which are difficult to understand because they involve an enormous quantity of matter and occur in huge volumes. The most important of these atmospheric events is the ongoing life of many animal and vegetable species.

The fundamental macroscopic proportions that define a fluid's properties are its composition, pressure, and temperature.

Earth's atmosphere is essentially composed of Nitrogen (78%), Oxygen (21%), and a multitude of other gases which, although a small percentage, are also very important for our life.

Because of human existence the chemical composition of the lower strata of the atmosphere (the Tropopause)

Gas	Symbol	%
Nitrogen	N_2	78.1
Oxygen	O_2	20.9
Argon	Ar	0.93
Carbon Dioxide	CO_2	0.033
Neon	Ne	trace
Helium	He	trace
Other Gasses		trace

Table 4.1: *Principal components of the Earth's atmosphere.*

varies a great deal. On a regional scale its composition is greatly affected by the presence of industries, power stations, traffic, air conditioning in buildings, etc. In many regions of industrialized nations, the concentration of certain gases is so high that the atmosphere cannot remove them all, thus creating an accumulation of gases, like CO_2, methane, chlorofluorocarbons, nitrogen oxides, etc.

Carbon Dioxide (CO_2) is an extremely important gas, because, along with water vapor, it is one of the most important causes, good and bad, of the Greenhouse Effect. Unlike Nitrogen and Oxygen, which allow received and emitted solar radiation to penetrate the atmosphere, CO_2 blocks the radiation emitted by the earth, thus causing an increase in the temperature of the atmosphere. Clearly this effect is extremely important for life on earth.

In many places on Earth human activity has produced large concentrations of industries which consequently produce large quantities of CO_2.

As we've seen, one phenomenon that could control the amount of CO_2 released into the atmosphere is chlorophyll photosynthesis. But deforestation (due to the increased demand for land), and forest fires (which produce a lot of CO_2), continually diminish its benefits. An increase in carbon dioxide causes an increase in the greenhouse effect, with the result, it is believed, that the temperature of the atmosphere could rise "too much" making life on Earth problematical!

Reducing the amount of CO_2 produced is a very significant current problem, and the subject of much discussion among politicians, scientists, and world populations. An initial step forward was made with the signing of the Kyoto protocol in 1997 which imposed limits on CO_2 emissions on signatory countries. Since, each year, the quantity of CO_2 produced continues to increase, other similar "global" conferences have been held, and other agreements for its reduction have been reached. Unfortunately, these agreements are not well respected, are often violated, and many times not signed by important nations, some of them great producers of CO_2.

In subsequent chapters we will consider in greater detail how the Earth-system, in the millennia before the industrial revolution, was able to control and maintain a low and constant level of CO_2, and how, in the last century it is no longer capable of doing so.

The other gases which contribute to the greenhouse effect are Carbon Monoxide, Sulphur Oxides, Nitrogen Oxides, and Methane. They are all also linked to human activity.

An increase in greenhouse gases in the atmosphere has another undesirable effect: acid rain. The acidity of a water solution is measured in pH: if the pH level is less than 7, the solution is acidic; if it's 7 it is neutral; if it's greater than 7 it is alkaline or base. Distilled water, for example, has a pH of 7. In the atmosphere, when water molecules react with CO_2 and other greenhouse gases they form carbonic acid, sulphuric acid, nitric acid, etc. Rain in unpolluted atmosphere has a pH level between 5 and 6.5; but if a high concentration of greenhouse gases is present (as in highly industrialized zones), the acidity of the rain will increase; if its pH level is less than 5, we get so called acid rain. Acid rain (including hail, snow, fog, etc.) deposits acids on both land and water, with devastating consequences for vegetation and marine life, and ultimately for human health.

The ratio of gases in the Atmosphere is more or less constant up to a height of about 180 km (112 mi). The proportions shown in Table 4.1 make up the so-called dry atmosphere. Luckily for us - otherwise we could not survive - a small amount of water vapor (from 0 -7%), which renders the air more or less humid, is also present in this mixture. Humidity is a physical measurement which describes the amount of water vapor in a given volume of atmosphere. Humidity is not constant, but varies in relation to place, time of day, season, temperature, altitude etc. The maximum amount of humidity that can be present in the atmosphere at a particular temperature occurs when water in liquid form is also present. At that point we say that the atmosphere is saturated and equals 100% relative humidity. This condition occurs in fog or when it rains. As stated, water vapor, like CO_2, is another very important greenhouse gas. As we will see later, the atmosphere will at times contain other gases and particles (called aerosols), produced by volcanos, fires, industrial pollutants, human habitation, etc.

Variations in its composition also greatly affect the atmosphere's ability to filter electromagnetic radiation, increasing or decreasing its ability to absorb or reflect certain wavelengths. Think for example about what happens to light when it passes through fog.

The atmosphere rises to a height of approximately 600 kilometers (370 mi), but its density (the quantity of air contained in 1 cubic meter) decreases as we get further away from the earth's surface. This decrease is very rapid; so, for example, the quantity of air in 1 cubic meter at an altitude of 3000 meters (1.8 mi) is about half what it is at sea level. Air density at the top of mount Everest is so low that only very few, well trained, people are able to stay there for more than a few minutes without oxygen tanks - and that's not even considering the difficulties in getting there.

The weight of a one square meter column of air, extending from the earth's surface to the top of the atmosphere, is called atmospheric pressure, and is measured in Pascal,[19] symbol: Pa. At sea level normal atmospheric pressure is approximately 100,000 Pa, or the pressure exercised by a mass weighing about 10,000 kg (10 tons) on one square meter. We say that atmospheric pressure is high if it is more than 1000 hPa (hecto Pascals), and low if it is less.

The general rule for the movement of air in the atmosphere is that air masses travel from high pressure zones to low pressure zones.

Figure 4.1 shows the progression of atmospheric pressure in relation to altitude.

Since its thermal properties vary significantly with altitude the atmospheric "skin" has been divided into layers: Troposphere, Stratosphere, Mesosphere and Thermosphere. [See Fig. 4.2]

The Troposphere, meaning sphere where changes occur, is the lowest region of the atmosphere. It extends to an altitude of about 15 km (9.3 mi). Since the Troposphere contains almost all of the total air mass in the atmosphere (about 80%), almost all meteorological phenomena occur in it. The temperature in this layer decreases about 6°C (42°F) with every 1000 m of altitude. At a height of 15 km air temperature is about -55°C (-67°F).

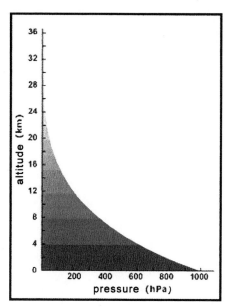

Figure 4.1: *Typical change in atmospheric pressure in relation to altitude. Note that essentially all of the atmosphere is located below 30 kilometers.*

The Stratosphere. In this layer, which extends to a height of about 50 km (31 mi), is found the greatest concentration of Ozone. Because of the presence of Ozone the temperature in the Stratosphere increases steadily with height to about -3°C (26°F). By absorbing Ultraviolet solar radiation Ozone not only acts as a virtual shield for life on Earth, but also heats up, and this is the reason for the increase in the temperature of the Atmosphere.

The Mesosphere is above the Stratosphere and extends to a height of about 80 km (50 mi). The temperature in this layer starts to decrease again arriving at about -90°C (-130° F).

The next layer is the so called **Thermosphere** which we believe extends to an altitude of about 600 km (370 mi). Because of the effect of solar radiation, the few atoms present in this layer are ionized and extremely energized, just as if they existed at high temperature. Satellites studying the Earth are located in this layer, in orbits at heights between 200 and 1000 km (125 and 625 mi). Although the density of the air in the Thermosphere is very low, it is still sufficient to slow down the speed of these satellites over the course of a few months, requiring a thrust from small on board boosters in order for them not to fall back to Earth.

4.3 - Aerosols in the Atmosphere

Aerosols are very minute solid or liquid particles suspended in the atmosphere, mainly in the Troposphere. These particles are almost always produced by natural phenomena, such as volcanic eruptions, desert dust storms, forest fires, or saline spray from the oceans. [See Fig. 4.3] These particles can vary in size from fractions of microns to hundreds of microns (1 micron is one thousandth of a millimeter!).

It is estimated that human activity is currently contributing

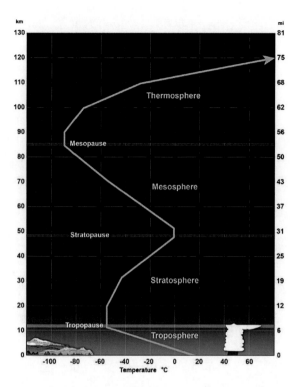

Figure 4.2: *Source NOOA. Variation in atmospheric temperature in relation to altitude, and the subdivision of the Atmosphere into Troposphere, Stratosphere, Mesosphere, and Thermosphere.*

to an increase in atmospheric aerosols by about 10%. Since aerosols originating from human activity are produced by industrial activity (factories, power stations, etc.), and by activities that make our lives more comfortable (chimney smoke, automobile exhaust, etc.), the major concentration of aerosols is found in the northern – most industrialized – hemisphere. Aerosols have multiple effects on human lives, ranging from allergies to cancer. Many of their toxic effects are inversely proportional to their size. In fact when particles are smaller than 10μm (these are called PM10, while smaller ones are called PM2), they are absorbed by the human body, causing damage that can be lethal. Therefore, the concern about decreasing this enormous quantity of material being emitted into the atmosphere on a daily basis, especially in large cities, is quite understandable.

In addition to these "local" effects, aerosols also have very important effects on the Earth's climate. Since many types of aerosols reflect sunlight back into space, they directly cause a decrease in the amount of solar energy that reaches the Earth's surface, causing it to cool. Under natural circumstances, aerosols form thin layers in the troposphere, which are usually eliminated by rain in about a week.

Figure 4.3: *Source, NASA. July 3, 2003. MODIS-Aqua satellite. Sand from the Sahara desert crossing the Mediterranean heading for northern Italy.*

An important source of aerosols is volcanic eruptions. These produce great quantities of aerosols together with large amounts of gases, particularly Sulfur Dioxide. When this gas combines with water vapor in the atmosphere, it produces Sulfuric Acid, an acid that causes grave damage to vegetation when it falls down to earth as rain (acid rain) or snow.

In 1991, in the Philippines, the volcano Pinatubo ejected a large quantity of aerosols directly into the Stratosphere. [Fig. 4.4] As we've seen, rain clouds do not normally form in the stratosphere, so that this enormous amount of dust had no way to return to earth in a short period of time. The effect was that we had a cooler than average summer (and splendid sunsets).

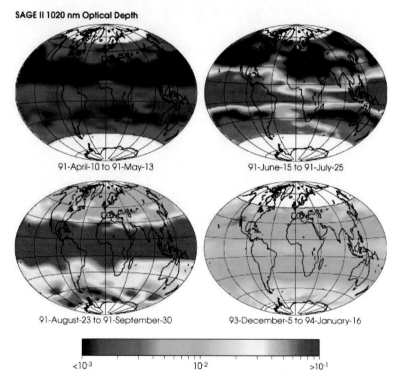

Figure 4.4: *Source, NASA. These images were taken by instruments on board the NASA SAGEII mission, and represent the quantity of aerosols in the Earth's atmosphere. The black areas indicate the lowest quantity, the red areas indicate quantities at least one hundred times higher. The image at top left shows aerosol levels before the eruption in June 1991 of the Pinatubo volcano (Philippines). At top right we see aerosol levels right after the eruption, at bottom left about two months later, and at bottom right a year and a half later. Aerosols produced by this eruption were detected more than ten years later.*

Perhaps the largest volcanic eruption for which we have sufficient scientific data was the eruption of Krakatoa in 1883. This volcano is located, or should we say was located on an island in the Indonesian archipelago, between Sumatra and Java. The explosion of Krakatoa was so sudden and devastating that only one third of the original small volcanic island remained. The ash ejected by the volcano reached ships almost four thousand miles away. The dust completely obscured the sky near the volcano for more than a day. It is thought that the volcanic ash emitted into the atmosphere screened so much solar radiation that the surface temperature of the Earth went down by about 1°C (34°F) for a year or two. The dust, which probably reached a height of more than thirty miles, as it filtered solar radiation, produced strange optical effects: for example, for some months violent red sunsets were visible in many parts of the world. But the most devastating effect of the eruption of Krakatoa was the death of more than 36,000 people. These deaths occurred not because of aerosols, but because of the gigantic wave (tsunami) produced when an enormous quantity of volcanic material fell into the sea.

4.4 - Energy transfer and clouds.

If in a physical system all phenomena were "natural," that is without "external" intervention, they would evolve over time in such a way that all the components of the system would have the same characteristics. This principle is demonstrated when we observe the way a warm body behaves when it comes in contact with bodies having a lower temperature. The warm body will give up some of its energy to surrounding colder bodies, until the whole system reaches the same temperature, called the equilibrium temperature, somewhere between the highest one and the lowest one. In other words, "Nature" tends to balance energy among the different parts of a system, redistributing it with "equanimity."

This fundamental law of Nature should give us pause to think!

In fluids, such as water and the atmosphere, this distribution is the key to understanding many phenomena. The basic components of the transfer of energy from more energetic to less energetic fluids are: convection, evaporation, and condensation.

Within a fluid, convection is the movement of warm particles toward colder zones, where they transfer part of their energy to cold particles. In the case of atmospheric fluid, subject to the Earth's gravitational force, the warmer, less dense part, rises, while the colder part falls. This creates a circular movement of rising warm particles (like air, water vapor, smoke, aerosols, etc.) and falling cold particles.

The direct result of convection motion is the movement of large air masses within the Earth's atmosphere, which generate winds and ocean currents.

The inclination of the Earth's axis, and the fact that the Earth is a sphere, together cause the alternation of the seasons, differences in the length of night and day, and a varying distribution of solar energy on its surface. This means that in the periods when the difference in temperature between polar regions and equatorial regions is highest (winter) there are strong air currents and ocean currents. These move from the equator toward the poles, from West to East in the Northern hemisphere and from East to West in the Southern Hemisphere. This phenomenon causes a global redistribution of energy within Earth's fluids.

As we've stated earlier, the transport and redistribution of energy particularly in the atmosphere, is not limited to convection, but also greatly depends on two other phenomena: the evaporation and condensation of water.

Water, i.e. the amalgam of H_2O molecules, can exist - depending on temperature and pressure - in three phases or states: solid (ice), liquid (water), gas (water vapor). At normal pressure, like at sea level, the solid state is reached at a temperature below 0°C (32°F); the liquid state at temperatures below 100°C (212°F); and the gaseous state at temperature above 100°C (212°F).

The process by which a liquid becomes gas is called evaporation. Water evaporates at all temperatures, but maximum evaporation is reached at 100°C. This is the temperature at which, at atmospheric pressure, a liquid boils and becomes gas. Energy is required to change water from one state into another, and also for heating it within the same state. 100 calories are needed to heat one gram of water from 0°C to 100°C, and more than five times that much (exactly 539 calories at atmospheric pressure) to change that same gram into gas. The amount of energy needed to change a liquid into gas or from a gas into liquid is called the latent heat of evaporation or condensation. In evaporation the energy needed to heat the liquid must be provided externally (thus cooling the external environment), while in condensation the water vapor transfers heat to the outside (thus heating the external environment).

By the process of convection large quantities of water vapor travel upwards from the surface of oceans, lakes, rivers, and humid land masses, to mix with large warm air masses. Since the temperature of the troposphere gradually diminishes as we move away from the Earth's surface (about 6°C (43°F) every thousand meters), warm water vapor[20] cools as it moves upward and condenses into minute water droplets, or sometimes into ice crystals. When water vapor (gas) condenses into droplets (liquid) it releases all the energy it had absorbed while becoming water vapor. The aggregation of these tiny particles of water or ice causes the formation of clouds. If these water droplets exceed certain sizes they tend to fall in the form of rain, or at times, hail.

As we've said earlier, clouds are found in the atmospheric layer closest to the earth, essentially in the troposphere. The greatest concentration of cloud formations is found in tropical zones where, since ocean temperatures are higher, water evaporation is more intense. Great cloud formations come together in zones with lower atmospheric pressure forming spiral cloud systems or cyclonic areas. These spin counterclockwise around a Low Pressure center; moving from West to East in the Northern Hemisphere and vice versa in the Southern Hemisphere. [Fig. 4.5]

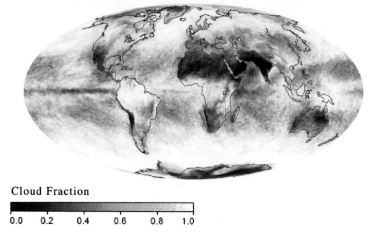

Figure 4.5: *Source, NASA. Average cloud cover over the entire planet in June, 2019. The map was created by the MODIS instrument on board the NASA-Terra satellite.*

From an energy standpoint, the formation of clouds - i.e. the absorption of energy from the oceans by evaporation, its transport by convection, and its release into the troposphere by condensation - is a perfect mechanism which, using water vapor as a transport vehicle, transfers energy from the Earth's surface to the troposphere, and then after discharging it, returns it to Earth. Due to the enormous size of cloud formations, this atmospheric energy transfer occurs on a global scale.

In addition to this global phenomenon there are many smaller energy redistribution systems, such as storms, hurricanes, and typhoons. The last two in particular develop in tropical waters and accumulate huge quantities of energy which they then discharge on land where they dissipate in the space of a day or so because they are no longer fed. These systems bring with them very violent winds and rains capable of causing great material and human damage. [Fig. 4.6]

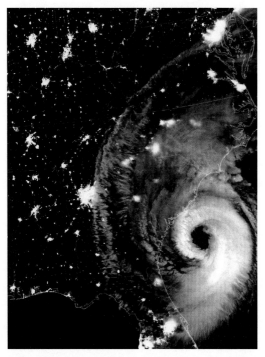

Figure 4.6: *Source, NASA. In 2019, After devastating the Bahamas and causing severe damage in Florida and Georgia, Hurricane Dorian gained more energy and struck the South Carolina coast with extremely violent winds and heavy rains, causing tremendous damage and numerous deaths. These nighttime images taken by the Suomi satellite in the visible and infrared ranges show Dorian as it hit the coast at 3:42 am (07:42 UTC) the night of September 5, 2019.*

Although each cloud is unique, most can be classified into two large families: Stratus and Cirrus. These two families then can be subdivided into other types according to their height, configuration, etc.

Stratus clouds are found in the lower levels of the atmosphere, up to about 3000 m. (1.86 mi), and are generally composed of water droplets. Fog is also a part of this cloud family.

Cirrus clouds, on the other hand, form at higher elevations, above 6000 m. (3.7 mi), and are thus composed of ice crystals.

In addition to being atmospheric water reservoirs, clouds interact with solar radiation (consider for example their color) in a very important way: they serve as giant sun shields controlling the temperature of the Earth's surface and regulating the amount of solar light which reaches it. The amount of regulation varies according to the types of clouds. [See Fig. 4.7] Stratus clouds are powerful reflectors of solar radiation back into space. In effect, they shield the Earth and make it cooler. Since Cirrus clouds are made up predominantly of ice crystals, they are very transparent and allow a large amount of sunlight to pass through. However, they do not allow the heat emitted by the Earth to pass through, thereby increasing the greenhouse effect and heating the Earth's surface.

A less common category of clouds, but one with unique thermal behavior, are the so-called Cumulus clouds. These clouds form vertically and reach a height of about 10,000 m. (6.2 mi). At lower heights their behavior is similar to that of Stratus clouds, at higher levels to that of Cirrus clouds. That is, they reflect solar radiation upwards and the Earth's infrared emissions downward. In other words, they behave like an excellent thermal insulator.

On a global level, clouds tend to cool down the Earth. Since they cover about 60% of the surface of our planet, we can see why studying cloud types and global cloud cover is so important to our understanding of the Earth's climate.

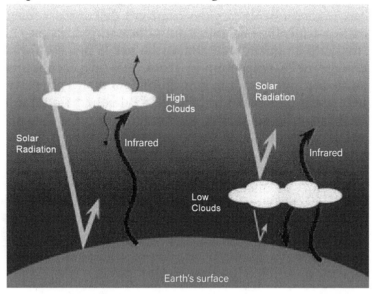

Figure 4.7: *Difference in behavior of Cirrus clouds (left) and Stratus clouds (right) with respect to solar radiation.*

4.5 - The Hole in the Ozone Layer

The subtitle of this section could be: The story of the errant Oxygen atom. In fact, the story of Ozone is closely tied to that of Oxygen atoms wandering around in the atmosphere. Ozone is one of the components of the atmosphere but is present in such minute percentages that in Table 4.1 it is included under "other Gases." (see pg. 69)

Ozone is a toxic gas which causes severe damage to lung tissues if breathed by humans. It is also highly toxic to plants. Fortunately, it is found in very small quantities in the Troposphere. It is however found in much higher concentrations in the Stratosphere - in a layer between 10 and 40 km., with a maximum concentration at about 25 km.

The Ozone layer in the Stratosphere is fundamental to life on Earth. It acts just like a protective shield against the sun's ultraviolet radiation. If UV radiation were not blocked by the Ozone layer, not only would we receive more solar energy on Earth, but would be more susceptible to skin cancer, cataracts, and many other illnesses. So we find that Ozone has a dual personality: Bad gas in the Troposphere, Good gas in the Stratosphere.

Ozone is formed indirectly by high energy solar UV radiation. UV radiation[21] is invisible to the naked eye and is made up of high energy photons, which explains their danger. What follows is the story of Ozone, and the story of the errant Oxygen atom.

Ozone (O_3) is a molecule consisting of 3 Oxygen atoms unequally bound to one another. The first two atoms form the Oxygen molecule, the third however is like an unwelcome guest whom we can't politely get rid of, but will do so the first chance we get.

The Oxygen molecule (O_2), formed by 2 Oxygen atoms, is a very stable molecule because the 2 Oxygen atoms are very tightly bound together. When an Oxygen molecule is struck by a high energy UV photon - i.e. by more energy than that which binds the 2 atoms - it breaks apart freeing the 2 Oxygen atoms.

$$O_2 + \text{high energy UV solar photons} = O + O$$

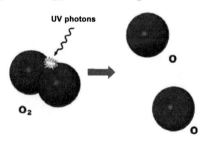

Figure 4.8: *The splitting of an oxygen molecule when struck by a high energy UV photon.*

Fortunately, the quantity of molecular Oxygen in the Stratosphere is such that all the UV photons striking it are absorbed and none reach the Troposphere. The Oxygen atoms produced by the reaction above, having absorbed all the extra energy not needed to break apart the O_2 molecule, now become very energized and begin wandering around in the Stratosphere until they encounter two Oxygen molecules; they latch onto them and form two Ozone molecules, O_3.

$2(O + O_2) = 2 O_3$

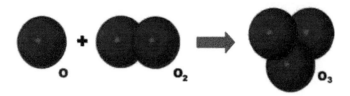

Figure 4.9 *Reconstruction of an ozone molecule.*

O_3 is a very unstable molecule because the third Oxygen atom has a very weak bond with the other two. This molecule is therefore easily destroyed by low energy UV rays (compared to those that break up the O_2 molecule).

The Oxygen molecule's job of absorbing high energy UV rays is now performed by Ozone with respect to low energy UV rays!

$$O_3 + \text{low energy UV solar photons} = O_2 + O$$

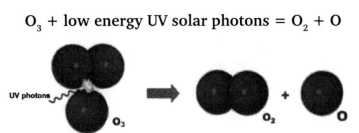

Figure 4.10: *The splitting of an ozone molecule by low energy UV rays.*

The Oxygen atom thus finds itself once again wandering around alone in the Stratosphere. But its solitude is short lived, because it soon recombines with an Oxygen molecule to reform a molecule of Ozone. The cycle of reactions just described begins again! This is known as the Ozone Cycle.

The high instability of the Oxygen atom and the Ozone molecule - that is their ability to combine easily with other atoms[22] - means that the preceding process is not the only one in which the Ozone molecule is destroyed. In fact various other gases exist naturally in the Stratosphere.[23] These are formed by compounds of Chlorine (Cl), Nitrogen (N), Bromine (Br) etc., whose molecules easily recombine with O and O_3 causing a decrease in Ozone concentration and in oxygen atoms.

Variations in Ozone concentrations in the Stratosphere also occur for various other reasons, such as the change in seasons, the effect of winds, changes in the intensity of solar radiation, etc.

Volcanic eruptions also cause reductions in Ozone concentrations because they inject great quantities of gases and aerosols into the atmosphere. But these reductions are recovered naturally within a year or two.

Although subject to minor variations, this natural process of reduction and creation of Ozone is such that the total amount of Ozone in the atmosphere is on average pretty much constant.

This delicate equilibrium has survived until the second half of the twentieth century. But since the early 1970's there is scientific evidence that the Ozone layer in the Stratosphere is becoming thinner and in certain areas is an outright hole.

We have discovered that this thinning is caused by the large increase in the concentration of Chlorofluorocarbons (CFCs) emitted into the atmosphere.[24] CFC molecules are composed of atoms of Chlorine, Fluoride, and Carbon, held together by a very strong chemical bond. These gases, which are present naturally in very small percentages in the atmosphere, are instead produced in massive quantities in various industrial processes. In fact CFCs are greatly utilized as refrigerants, in solvents, in detergents, in products used to fight fires, and among other things are used as propellants in spray deodorants, aerosol cans, etc.

Since they are very stable, CFC molecules do not interact with other molecules. And although they are heavier than air, when pushed by ascending air currents they can easily reach the Stratosphere, where they can be struck by high energy UV photons. Only these photons are capable of breaking up CFC molecules; they break the weakest bond, the one holding the Chlorine atoms together. [Fig. 4.11]

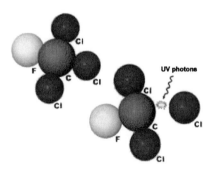

Figure 4.11: *A CFC molecule is composed of three chlorine atoms Cl, one carbon atom C, and one fluorine atom F. When it is struck by a high energy UV photon it can lose one of its chlorine atoms.*

When a free Chlorine atom approaches the very unstable Ozone molecule it strips away one Oxygen atom and forms a molecule of Oxygen and a molecule of Chlorine Monoxide, (ClO). [Fig. 4.12]

The ClO molecule then quickly captures a free Oxygen atom thus forming a new Oxygen molecule and again freeing the Chlorine atom to search out and destroy another Ozone molecule. [Fig. 4.13]

$$Cl+O_3 = ClO+O_2$$

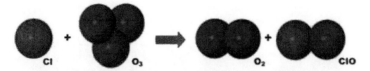

Figure 4.12: *A CFC molecule strips an oxygen atom from an ozone molecule, forming an oxygen molecule and a chlorine monoxide molecule CLO.*

Fortunately, Chlorine does not last long in the atmosphere, or it would destroy all of the Ozone. But while it remains in the atmosphere (approximately one week) a Chlorine atom destroys, on average, about 100,000 molecules of Ozone.

The reduction of Ozone in the Stratosphere is particularly great over both Poles in the winter months. Because of the lack of sunlight, the temperature of the Stratosphere in these zones is very low -less than -80°C (-112°F). This favors the formation of stratospheric clouds, called Polar Stratospheric Clouds. These clouds are composed of ice crystals on which CFCs deposit themselves and release Chlorine.

$$O+ClO= O_2 + Cl$$

Figure 4.13: *A ClO molecule quickly captures an atom of oxygen and reforms an oxygen molecule and a chlorine atom which is ready to destroy another ozone molecule.*

Special instruments on board satellites studying the Earth's atmosphere have identified an enormous hole in the Ozone layer above the South Pole almost as large as the surface of North America. Lethal UV rays can penetrate the atmosphere through this hole and reach us. [Fig. 4.14]

Figure 4.14: *Source, NASA-GSFC. A succession of images, taken by the TOMS instrument, which show the increase in the Ozone hole from 1979 to 1999. The color blue represents areas where the Ozone layer is very thin. The range from green to yellow represents areas where it becomes progressively thicker.*

The dimensions of this hole vary, depending on the season, industrial production of CFCs, volcanic eruptions, etc. How can we repair the hole in the Ozone? There is only one way: by decreasing the production of products that contain high levels of CFCs. In 1987, in Montreal, almost all the world's nations signed a protocol limiting the production of CFCs. The protocol was made more stringent in 1992, and again in 1996, by obligating all industrialized nations to suspend the production of CFCs. It is estimated that if the production of CFCs is not suspended, by 2050 the number of persons affected by cancer in the US will increase by 33,000 cases per year, and in Europe by 14,000 cases.

It is believed that the commitments just mentioned have brought about a shrinkage of the hole in the ozone layer, and therefore less UV rays reaching the earth. In the last few years, however, due to the discovery of other gases which can destroy ozone molecules, the situation has become more precarious. Let's hope we can reverse this trend: it's only up to us. [Fig. 4.15]

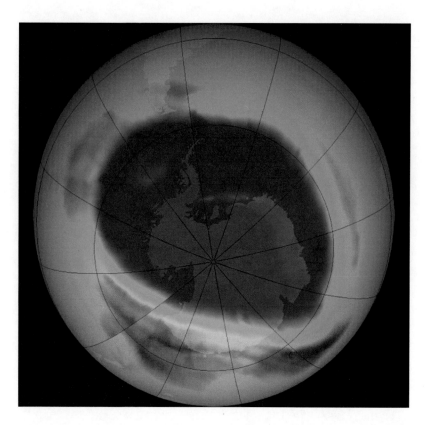

Figure 4.15: : *Source, NASA-GSFC. An image of the hole in the ozone layer taken by the OMI instrument aboard the AURA satellite on September 17, 2014. The hole seems to have shrunk after the prohibition of CFCs. The color codes are the same as in Fig. 4.14.*

5

WATER FOR LIVING

5.1 - El Niño? Who is this guy.

We were spending all our time with the Doctor and drinking a mixture of water and salty minerals from these special containers. "Water," said the teutonic Doctor, "is the most important thing for living. Where does it come from? A mystery! Some theories maintain that water was transported to Earth about four billion years ago by comets or asteroids that struck the planet."

"We need water because we essentially are water."

"Just as the ancient mariners stocked fresh water on their ships, so must we too store water in our space station, and we must be very careful not to waste any. But we can't count, like they did, on replenishing water from rain or on land, asssuming they found any. Here we must use as little as possible, and we must recycle most of what we use."

I pulled the straw I was drinking from out of my mouth, but Caro, who still hadn't figured out where this lecture was headed, wanted to find out more. The doctor went on: "The body gets rid of all the water you drink, when you sweat, when you pee, and when you make number two. On Earth these waste products, rich in water, remain on the planet and one way or another are recovered. The same thing, more or less, has to happen on this space station. We have machines which recycle the water our body gets rid of." And then, like all these crafty geniuses, he went on: "For example the beverage we're consuming now could very well be the recycled urine of the station's occupants." You should have seen the look on Caro's face when she sprayed out everything she had in her mouth. The only person laughing was the doctor. The two of us decided to stop drinking all together for the rest of the day. But then the Commander started in with her motherly encouragement, and when we saw that everybody else was drinking, we started to drink too.

"On Earth we don't lose even one drop of water; whatever leaves as water vapor returns as rain or ice. What goes into the ground winds up in rivers, rivers flow into the oceans, etc. etc.

"The water that has existed on Earth for thousands and thousands of years is always the same because it is constantly being recycled. It's possible that a rain drop falling to the ground is the same drop that a person drank thousands of years ago. It would seem then that humanity should not have a water problem. But it does, and it's a big one. Just as in the past there have been more than a few wars waged over water, we fear that the same may happen in the future.

"The human body needs 'potable' water. The risk we face is that there won't be enough; either because we waste it, or because we pollute it. So, there is the possibility that we will be without it for a period of time, longer than what the Earth needs to purify it again. Under those circumstances the situation would truly be dramatic. We can go a few days without eating - and some of us should probably try it - but we can't go without drinking for very long, we would die for sure.

"Rains are our main source of distilled water. But rains aren't always benign. Sometimes they arrive with such violence, as in tempests or hurricanes, that they cause enormous damage.

"While we're on the subject of bad weather, do you know about El Niño?"

We had no clue...

"Some years in Peru the temperature of the Ocean rises a degree or two higher than normal. When that happens very violent rains strike the coast and the interior of the country. These rains are a great disaster in terms of human life and material and economic destruction.

"One of the most serious economic losses is the sudden decrease in fish stocks in the area where the sea is usually very rich in anchovies. At the same time, almost as compensation for the damage, we witness some unexpected natural plant growth in the Peruvian desert, and an abundance of fish and shrimp in lakes newly formed by the heavy rains. Peruvian fishermen have named this phenomenon (which strikes not only their country but a much larger area on Earth) El Niño,[25] or the Child. Fortunately for the fishermen, the following year Ocean temperatures decrease, anchovies become plentiful again, and the desert is once again arid. They call this phenomenon La Niña (the girl child)."

While the doctor was telling us about all these important events, all I could think about was Santa Claus sitting in a giant shell pulled all over the place by millions of anchovies.

"Have you ever thought about the amazing properties of water? It's a great cleaner and solvent, especially good a dissolving mineral salts; it's colorless and transparent; it absorbs infrared radiation, and therefore absorbs and then reflects "the heat" of any body it touches; it can become solid or gas within a temperature range of only 100°C (212°F), and with these changes is able to transfer energy; it is capable of climbing upwards against gravity, along capillary vessels and reach the highest branches of plants; it is tasteless when distilled, but luckily for us, fresh and salt water both have taste.

"Water is everything for our life. If we could analyze your 110 pounds of body weight Caroline" - it was her day - *"we would observe the following: around ten and a half gallons of water (close to 85 pounds), 15 and a half pounds of stones (calcium), a few coals (carbon), a few other elements, and a tiny bit of grey matter."*

Caroline did not appreciate the reference to her weight, but the Doctor had made his point!

5.2 - The general properties of water

Water, like the atmosphere, is a fluid which is indispensable to life. It makes up from 50 to 90% of all forms of life on Earth. Our bodies consist largely of water, about 80%. And more than 70% of the surface of the Earth is covered by water. It is estimated that the total quantity of water on Earth amounts to approximately 1500 billion kilograms (1.4 billion tons),[26] a good 300 times the total amount of fluid in the atmosphere.

Let's remember that we didn't start out as fish purely by chance.

Because water can exist as a solid, liquid, or gas, it has, among many amazing characteristics, one that is unique: it requires a lot of energy to heat[27] and needs a lot of time to cool.[28] (If you've ever gone swimming at night, you'll understand). This characteristic means that the oceans are immense reservoirs of heat which they distribute from temperate to colder zones. If we add evaporation and condensation to these characteristics, we see why water is the principal fluid needed for the stockpiling and transfer of energy on planet Earth.

All of these amazing characteristics depend on water's strange molecular structure, formed by one Oxygen atom two Hydrogen atoms. The Oxygen atom, which has a greater mass than the Hydrogen atoms, is very willing to host the others' electrons. This means that the electrons of the two Hydrogen atoms are fixed guests of the Oxygen atom.

Having been abandoned by their respective electrons the Hydrogen nuclei - positively charged - repel each other, so that the composition of the H_2O molecule is a triangle whose angle at the top is 105°. [Fig. 5.1] This molecule is not "well balanced." The Oxygen side is heavier and has a negative charge, the Hydrogen side is lighter and has a positive charge. Therefore, when a positively charged ion approaches a water molecule it will be attracted to the Oxygen side, and a negatively charged ion will be attracted to the Hydrogen side.

This happens with salts, which are formed by positive and negative ions united by a weak bond. When salts are placed in water they dissolve, because their positive ions are attracted by the water's negative pole (oxygen), while the negative ions are attracted by the positive pole (hydrogen).

Figure 5.1: *The H₂O molecule. The two Hydrogen nuclei, protons positively charged, repel each other. But since they are bound to the Oxygen atom by their respective electrons, they form a triangle with O at the apex and the two H at the base.*

Another incredible property of water is related to its density. Whether solid, liquid, or gas when a body cools its volume decreases. The opposite happens when its temperature increases. This is easily understandable because heating means providing energy to molecules, and cooling means removing energy from them. If molecules are more energized, they are more active and seek to occupy more space; if they are less energized they are less active and seek to occupy less space. A body's density equals its mass divided by its volume. Therefore, when a body is heated and its mass remains constant while its volume increases, its density will decrease. The opposite happens when a body is cooled.

This fact is only partially true for water. In fact, when the temperature of water decreases its density increases until we reach a temperature of 4°C. But as the temperature decreases further its density begins to decrease and its volume begins to increase. When the temperature of water reaches 0°C, it freezes and forms ice, whose density is approximately 10% less than water, and fortunately floats. We say fortunately because if ice were denser than water it would sink and accumulate at the bottom. If this were to happen in the oceans the sunken ice would trap all forms of life. Instead the floating ice, being an excellent thermal insulator, keeps the water below from freezing and allows all marine organisms living at depth to keep swimming around very peacefully. To say that all of this is incredible is probably an understatement.

Water is an excellent solvent. Therefore pure water found in nature always finds ways to "get dirty." In its constant flow, water dissolves almost everything in its path and becomes contaminated with lots of impurities, both organic and inorganic. This is why the amount of potable water at our disposal is a very small part of all the water found on Earth. In fact about 97% of all water is in the oceans, and the remaining 3% is divided almost equally between glaciers and underground. [Fig. 5.2]

Less than half of all underground water can be used by humans. To put it more simply: for every 100 liters of water found on Earth only about the equivalent of a half-liter bottle is fresh water, which can be potable if we don't pollute it. This knowledge is very disconcerting but also indispensable. In order to better understand the characteristics of water tied to our "life" we must examine its "quality" as determined by a number of indicators.

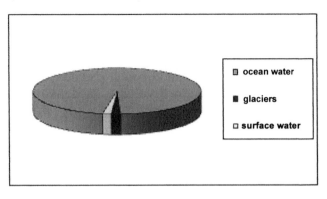

Figure 5.2: *Percentage distribution of water on Earth. Oceans (blue), Glaciers (red), Underground water (yellow).*

Transparency

This is the capacity of water to allow visible light to pass through it. The transparency of water is an indicator of its ability to foster chlorophyll synthesis, in other words its ability to promote the growth of marine plant life. The less transparent water is, the less sunlight can penetrate it; consequently less vegetable life will be found, and at lower depths. The transparency of water depends on three types of substances: (1) solid particles in suspension - inorganic materials transported by rivers (like mud), agricultural residues such as fertilizers, sediments on the ocean floor placed in suspension by waves or currents; (2) yellow substances: particles of organic material produced by the decomposition of vegetation, etc.; (3) phytoplankton: microscopic green plants that make up the first step in the marine feeding chain.

Of course, far from coastlines the substances which have the greatest effect on transparency are phytoplankton, while transparency near river deltas is greatly affected by the presence of solid particles in suspension, or by yellow substances.

Temperature

From a thermal standpoint, the Ocean is divided into two zones: one heated by the sun, and one not heated by the sun. [Fig. 5.3]

In the first zone water temperature depends on its exposure to solar radiation. This means that ocean temperature in tropical zones is higher than in temperate or polar regions. It fact it varies on average from about 30°C in the tropics to about 0°C at the poles.

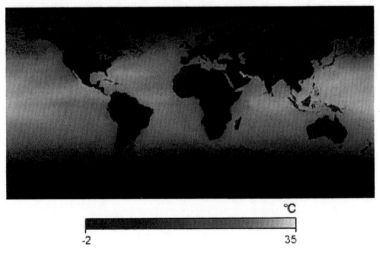

Figure 5.3: *Source, NASA Earth Observations (NEO). NASA Ocean Color Group. Ocean surface temperature in June, 2019. Data obtained by MODIS/Aqua satellite. Water temperatures range from -2°C (dark blue) to +35°C (white).*

Water temperature can be even higher near coastlines and near human discharges. The higher the temperature the greater the process of evaporation, and therefore the greater the amount of water and energy transferred to the atmosphere.

At greater depth, where, because of lower transparency less solar energy reaches, the temperature of the water is lower. Below the surface there is an area of demarcation where the temperature drops rapidly, then changes more gradually with depth, until it reaches a temperature of only a degree or so at the greatest depths. This layer of demarcation between the warm superficial zone and cold deep water is called the Thermocline. At the poles the thermocline is practically non existent because the surface temperature and the temperature at lower levels are more or less the same. In temperate latitudes the thermocline reaches its greatest depth at about 1000 meters. In tropical regions, where surface temperatures are very high, the thermocline is not very deep, about 100 meters, but is well defined, and behaves like an insulating material, in that it prevents the cold water below from mixing with the warm water at the surface.

The temperature of ocean water, both at the surface and at depth, can be lower than 0°C. This is due to its salinity and the pressure to which it is subjected: for every 10 meters of depth there is a corresponding increase in pressure of 1 atmosphere.

Marine life is greatly influenced by water temperature. Therefore, it is very important to have instruments which will allow us to continually monitor oceanic temperature on a global scale. The best place from which this can be done is space, from satellites orbiting the Earth.

Salinity
Because water is capable of dissolving almost

everything it encounters, the oceans contain great quantities of minerals. The most abundant are Chlorine and Sodium, whose combination, Sodium Chloride (NaCl), is more commonly known as Salt. [Table 5.1]

Salinity is the concentration of all minerals, essentially of all mineral salts, present in water. In other words, it is the ratio between the quantity of salt and the quantity of water in which it is dissolved. Mineral salts are always present in water, and their concentration strongly influences the development of life in it. On average the salt concentration of ocean water is approximately 3.5%, or about 35 grams of salt per one kilogram of water. In the water we drink it varies between 1 and 5 grams per liter.

Elements	% of mineral content
Chlorine	55.3
Sodium	30.8
Magnesium	3.7
Sulfer	2.6
Calcium	1.2
Potassium	1.1

Table 5.1: *the major components of ocean water.*

The salinity of water at the surface of the oceans varies with latitude and seasonal changes. [See Fig. 5.4] Generally in the middle latitudes - those between tropical and temperate zones - the salinity of water in the warmer seasons is higher than what it is in other zones. This is due to the increased rate of evaporation. Equatorial waters on the other hand are less salty because frequent rainfalls there discharge "distilled" water on the surface. Likewise at the poles, where fresh water is produced by melting ice. Salinity is also lower near the deltas of large rivers which deposit fresh water in the ocean. In ocean depths salinity is generally lower than on the surface, and changes very little.

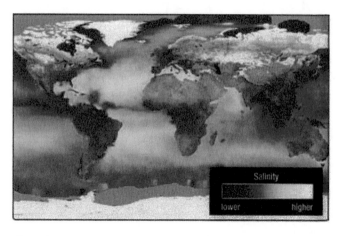

Figure 5.4: *Source, NASA-CONAE. Map of ocean salinity obtained from data provided by the JPL-Aquarious instrument aboard the Argentine mission SAC-D.*

Dissolved Oxygen

The key element essential for marine life is dissolved Oxygen. All marine life needs oxygen; but not the oxygen contained in water molecules. It needs the oxygen which, because of its constant movement, is dissolved in the water. The percentage of oxygen in the air is about 20%, in water it ranges from .001% to about .01%. Under .003% all forms of life disappear.

The presence of dissolved oxygen means that photosynthesis can occur, and phytoplankton and marine life can exist. Where we find dissolved oxygen, we also find bacteria which decomposes organic matter. Where we find phytoplankton and bacteria we find small fish, and where small fish are we also find big fish, and so on.

Water temperature is critical for the existence of marine life. Although it is true that more photosynthesis occurs in warm waters, it is also true that at higher temperatures water is less able to retain its dissolved oxygen. Therefore, when water is warmer, as happens in summer, we find less fish in the sea.

Acidity

The oceans are able to absorb a great quantity of Carbon Dioxide (CO_2) present in the atmosphere. This very positive effect also has a negative effect: as CO_2 dissolves in sea water it renders it more acidic. Consequently, the more CO_2 in the atmosphere, the more acidity in the oceans. It is estimated that since 1850 the pH level of the oceans has decreased from 8.25 to 8.14. Although a small decrease, this means that the oceans have become more acidic, and have consequently threatened the survival of places like coral reefs, of shellfish, and threatened marine life in general.

Calcium (Ca)

Water is a great solvent of Calcium Carbonate, the basic element in many rocks. Calcium is very important because of its capacity to lower the acid level of water.

Nutrients

A number of indispensable substances for marine biology, even if in small concentrations, can be found dissolved in water. Nitrogen (N), Phosphorus (P), Carbon (C), at times Iron (Fe), Silicon (Si) etc. Nitrogen can be found either as a dissolved gas or as Nitrate (NO_3), which comes from rainwater or the decomposition of organic matter. Too much Nitrogen or Phosphorous causes excessive growth of algae or mucilage. Carbon is very abundant in the air as Carbon Dioxide, and, like Oxygen and Nitrogen, is taken directly from the air and dissolved in water.

Phytoplankton

Phytoplankton are made up of microscopic green marine plants and are the basic element in the marine food chain. As with all plants, they need light, water and nutrients in order to grow. Therefore, they are found in abundance near the ocean surface (where more sunlight reaches and where Chlorophyll photosynthesis occurs more intensely), at river deltas (where the water is richer in nutrients), or where deep ocean currents touch the surface. Water rich in phytoplankton has a wonderful greenish hue due the presence of Chlorophyll. And wherever there are phytoplankton there are more fish and the water is therefore cleaner. [Fig. 5.5]

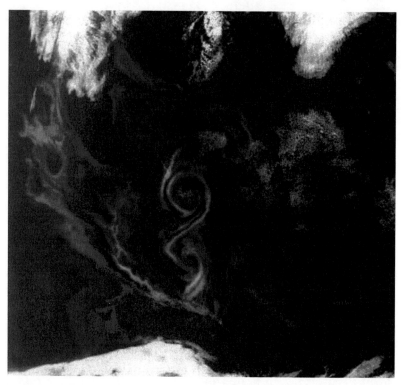

Figure 5.5: *Source, ESA. Meris/Envisat Jan.13, 2012. Summer phytoplankton growth in the southern Atlantic Ocean. It's very important to monitor the growth of these microorganisms, since climate changes may cause an increase in the growth of toxic microorganisms.*

Phytoplankton have an extremely important role in climate control. During photosynthesis they remove Carbon Dioxide from the sea and release Oxygen. Like all other plants they need carbohydrates to grow. Carbohydrates are essentially made of Carbon taken from the atmosphere and now present in the sea. Phytoplankton die after a day or two, their shells sink to the bottom of the ocean, and create an efficient dump site for the Carbon produced on Earth. On land, plants behave differently. When they burn or decompose, the Carbon they contain is once again discharged into the atmosphere. [Fig. 5.6]

As we can see, then, the more phytoplankton in the ocean the more Oxygen is produced, and the more Carbon Dioxide is removed. We estimate that marine plant life captures more than 1/3 of all the CO_2 on Earth and produces more than 50% of the Oxygen in the atmosphere.

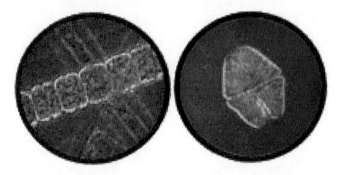

Figure 5.6: *Source, SeaWifs Project. Center and D.W. Coats. Different species of Phytoplankton. Although they have different forms, they all have a similar color due to chlorophyll.*

5.3 - Ocean currents

Water and the atmosphere are two fluids which, with minor differences, tend to behave in similar ways. We've seen how the evolution in time of the natural processes in a physical system like our Earth is governed by a very simple law: that of minimizing the energy differences which exist among the parts of the system. In fluids this law manifests itself in the convective motion of particles; that is the movement of particles from a warm area - where they have acquired energy - to a cold area where they "push" cold and less energized particles to warm areas. This constant flow carries warm particles to cold particles and cold particles to warm ones with the intent of establishing a thermal equilibrium between the two zones.

This is the engine that drives atmospheric currents or winds which travel from warm tropical zones toward cold polar zones. The same thing happens in the water, where oceanic currents - enormous quantities of warm water - travel from the Tropics toward the Poles. Due to the effect of the Earth's rotation, Ocean currents (as we've already seen with winds) travel clockwise in the Northern hemisphere and counterclockwise in the Southern hemisphere.

The two fluids, water and atmosphere, are separated by an interface which shares characteristics of each system. This interface is simply the surface boundary between the two fluids. Depending on how it is utilized, this is the boundary separating the oceans from the atmosphere or vice versa. Water and air exchange energy across this boundary in the form of heat or mechanical energy; or they exchange mass in the form of water vapor or rain.

The exchange of mechanical energy occurs due to an important difference between gases and liquids: the former can be compressed, the latter cannot. A direct consequence of the incompressibility of water is the formation of waves which form on its surface. When part of the wind's energy is transferred to the surface of the sea it responds by forming waves. Clearly the stronger the wind, that is the more energy it transfers, the higher the waves. To be more precise, the height of waves depends not only on the wind's strength but also on its persistence, and on the depth and length of the marine basin. For example, waves are shorter in lakes than they are in the Mediterranean, where they are shorter than in the open Ocean.

As the wind blows above the ocean it moves and stirs up the surface so as to form a layer of water - between 10 and 200 meters thick - of almost uniform temperature and salinity. In addition, constant winds guide masses of warm surface water toward the poles. Water moves very slowly in response to the action of the wind; in fact, ocean currents travel at just a few kilometers an hour. Because water has a high specific heat, a high thermal inertia,[29] and covers 70% of the Earth's surface, the oceans are gigantic reservoirs of calories, and ocean currents are enormous "conveyor belts" of energy. These enormous belts, tens of kilometers wide and as deep as 200 meters, are capable of transporting from the tropics to the poles about half of all the heat stored in the oceans. Part of the energy in the currents will be transferred along the way to the atmosphere, the rest will be given up to polar waters. When these surface currents cool, their density increases and they sink toward the bottom. At that point a current of deep cold water begins moving toward equatorial zones in the opposite direction to that

Figure 5.7: *Currents which transport energy in the oceans. The dotted arrows indicate deep water currents, the solid arrows warm surface currents.*

of surface currents. As they begin to arrive at the equator, after about a thousand years, they once again start their climb toward the surface in order to heat up and begin again a new trip on the "conveyor belt."

The Gulf of Mexico current, for example, carries warm water to Northern Europe and returns cold deep water toward the Equator. Thanks to this providential current Northern Europe's waters are on average 4°C warmer than those in other parts of the Earth at similar latitudes.

5.4 - Oscillations of the South Pacific and North Atlantic

The exchanges of energy between water and the atmosphere are fundamental in determining global climate, and anomalies in these exchanges can cause profound changes in it.

Trade winds are periodic winds which blow from high pressure zones, in sub-tropical regions, toward low pressure zones near the equator. In the southern hemisphere they blow from SE to NW and in the northern hemisphere from NE to SW. [Fig. 5.8]. South of the equator, in the Pacific

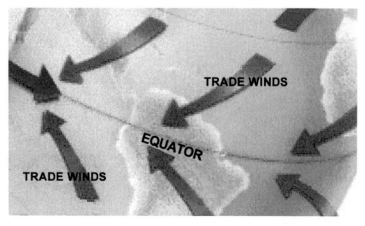

Figure 5.8: *The direction of Trade Winds in the two hemispheres.*

Ocean, trade winds travel from the South American coast (Equador, Peru) westward toward the Indonesian coasts. These winds also move these equatorial waters away from the Peruvian coast toward Indonesia and Australia, so that the western Pacific is about 50 cm higher than it is in the East.

As they pass over warm ocean waters these trade winds enrich themselves with water vapor which rises rapidly upward to form unstable cirrus cumulus clouds; these in turn produce violent rains in the middle of the Pacific Ocean. Once the air has cooled and lost its humidity it returns toward the East.

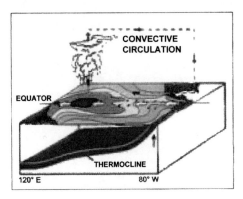

Figure 5.9: *Source, NASA. Normal conditions south of the Equator. High Thermocline, intense rainfalls in the Pacific Ocean between Indonesia and Peru.*

These are what we call "normal" conditions: in other words, a difference in temperature between Indonesia and Peru; waters which are warmer in the West than in the East; a sufficiently high Thermocline near Peru, where the coastal waters are colder and rich in fish; a relatively dry climate in the area, and desert like regions inland. Indonesia, on the other hand, is very humid and rainy. [Fig. 5.9]

At times the temperature of the ocean near the Peruvian coast rises slightly. When this happens, the Thermocline is deeper. The trade winds, now fed by a smaller difference in temperature, will be less intense and wetter, so that storm laden cirrus clouds will not form in Indonesia, but in the central Pacific. The winds returning East will be rich with water vapor which now will fall violently and abundantly on the Pacific coast of Peru and nearby countries. This phenomenon occurs every four to seven years during the winter months, hence the name El Niño, given to it by Peruvian fishermen, referring to Baby Jesus.

When this happens, the sea is poorer in fish, but so much rain falls - as much in a few days as in a whole year - that the deserts bloom and inland lakes become larger and rich in fish. These storms often bring with them much destruction and loss of life. In Indonesia, on the other hand, there will be high pressure and drought. [Fig. 5.10]

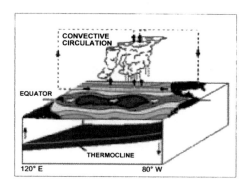

Figure 5.10: *Source, NASA. Conditions favorable to the formation of El Niño. Low Thermocline, intense rainfalls in Peru.*

El Niño is accompanied the following year by another phenomenon called La Niña. This time the Thermocline is shallower than under normal conditions, the Peruvian waters become colder than normal, the ocean richer in fish, and more rain will fall on Indonesia. [Figure 5.11]

This recurring phenomenon: the periodic heating and cooling of the tropical waters of the pacific near South America, and the corresponding cooling and heating of another area of the Pacific in Asia, extended also to global

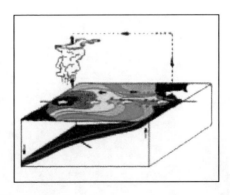

Figure 5.11: *Source, NASA. Conditions favorable to the formation of La Niña. Very high Thermocline, intense rainfalls in Indonesia.*

and atmospheric weather patterns, is called the Oscillation of the South Pacific.

Thanks to the help of very powerful computers it has been discovered that various areas of the Earth's surface are subject to similar meteorological oscillations. We say that these areas are "teleconnected". In other words, if for example the pressure at point A on earth increases, at point B very far away it will decrease.

This phenomenon, called "teleconnection" was first discovered between a location in Australia (Darwin) and another in Polynesia (Tahiti) more than 15,000 kilometers away. El Niño and La Niña are two aspects of teleconnection within the more general phenomenon of South Pacific Oscillation. Since there are areas very far from Peru and Indonesia which are teleconnected, a good understanding of the El Niño (or La Niña) phenomenon, or even better, the ability to forecast them, will enable us to predict what will happen in teleconnected areas. In fact during El Niño in the winter of 1997 [Fig 5.12] with the droughts in Indonesia and torrential rains in Peru, a great drought also occurred in Mexico and Central America, while the southern United States and Alaska experienced violent rainfalls and higher temperatures.

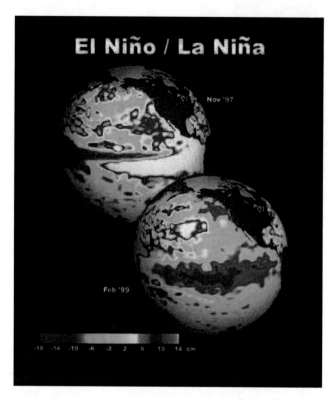

Figure 5.12: *Source, NASA-JPL-CNS. TOPEX/Poseidon is a French-American Mission which studies ocean height. The scale at bottom shows height and depth. In the top image (November 1997) we can see that the Pacific Ocean near South America is a few centimeters higher than in the East. This is true under El Niño conditions. In the bottom image we see conditions under La Niña, where the waters near the coast of Indonesia are higher than normal.*

These Pacific oscillations have also been discovered in the North Atlantic. Specifically teleconnection exists between what happens near the Azores and what happens in the area between Ireland and Greenland. It would seem therefore that what we've called anomalies in the interaction between Water and Atmosphere should instead be seen as beats or pulsations between two coupled systems exchanging energy with each other.

We do not yet have at our disposable theoretical models capable of explaining and accurately predicting the El Niño phenomenon. Some empirical models allow us to predict an El Niño a few months in advance. Recent theoretical models, developed on the hypothesis that El Niño is triggered by gravitational forces like tides, are very promising in their ability to accurately predict its occurrence.

5.5 - Fresh Water and Polar and Mountain Glaciers

As previously noted ocean water is rich in mineral salts, especially Sodium Chloride. Depending on its temperature a certain percentage of water will evaporate and pass into the atmosphere.

The process of evaporation has three principal consequences: in order to evaporate, one gram of water needs a great amount of energy - more than 500 calories (the same amount of energy needed to raise the temperature of 100 grams of water by 5 degrees) - which, of course, is removed from the ocean. The resulting water vapor is chemically pure. As it rises in the atmosphere water vapor cools and condenses, forming water droplets or ice crystals. During this process it transfers the energy it took from the ocean to the atmosphere. As droplets form, they capture everything they find in the air, like aerosols, chlorine, etc. This means that the rain or snow that falls to Earth is not pure, but contains various substances. Acid rain is one of the results.

The so-called fresh water that returns to Earth and forms rivers and lakes, rich in mineral salts and at times purified as it passes through the Earth's surface, is what we drink.

We must be careful to use this fresh water wisely, since it is such a minute portion of all the water on Earth. The victims of water shortages are innumerable. Wars are waged over water!

In order to sensitize public opinion to this problem - which is much graver than what we'd like to believe - the United Nations declared the decade 2005-2015 as the "International decade of Water for life."[30] [Fig. 5.13][31]

Figure 5.13: *"Water is not a privilege...it is a right." With this motto the international organization Green Cross has launched an important initiative to ask all governments to adopt a Global Convention for the Right to Water.*

A portion of water vapor falls to Earth as snow. Snow is a simple and very efficient way to conserve water. In temperate zones, as the temperature rises, snow melts and fills springs and rivers. But at mountaintops or in polar regions summer temperatures are not high enough to completely melt the snow, and as a result over the years it accumulates sufficiently to reach a sort of equilibrium. This snow surplus which doesn't melt forms glaciers. There are two types of glaciers: those that form on land with fresh water, and those that form from ocean salt water in the low temperatures reached in polar regions. But since it also snows in these regions, salt water glaciers will also contain ice made from fresh water.

The first type of glaciers includes all mountain glaciers, the glaciers that cover Greenland and lands near the North Pole, and those that cover the South Pole. These glaciers constitute the largest fresh water reservoir on Earth, more than 75% of all the available fresh water. The second type includes all the glaciers that form in the ocean, ie. all the glaciers at the North Pole, and part of the glaciers in the South Pole.

A big difference between the two types of glaciers is the effect each would have on sea levels if they were to melt. If fresh-water glaciers (those on land) melt they will cause a rise in sea levels; but if ocean glaciers dissolve, sea levels will not be affected.[32]

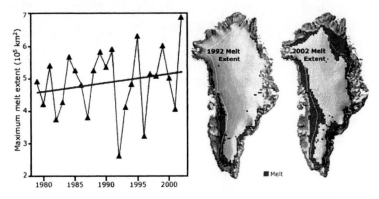

Figure 5.14: *Source, NASA. Change in surface area covered by glaciers in Greenland in the last decades. The graph at left, which shows changes in glacier melting between 1980 and 2000, demonstrates a constant increase in the melting of mountain glaciers.*

Snow and glaciers are extremely important for the maintenance of the energy balance on Earth; namely for the control of the amount of net energy we receive from the Sun. We know that snow and ice reflect light much like mirrors. In fact, they behave exactly like mirrors, by reflecting the solar radiation that reaches the surface back into the atmosphere.

Surfaces covered by rocks or vegetation behave quite differently. Therefore, if land glaciers were to decrease - causing a rise in sea levels and further exposure of land mass - less solar radiation would be reflected into the atmosphere, and more would be absorbed by the Earth's surface, which in turn would become warmer, causing even more melting of the glaciers. [See Fig. 5.14 for the situation in Greenland] All of this would be accompanied by a further increase in sea levels, with likely grave consequences for coastal cities (which are not few in number).

Sea levels can be "easily" monitored from space with an instrument called a radar-altimeter. By calculating the time a radar signal takes to travel from a satellite to the surface of the sea and back, this instrument can - with almost millimetric precision - measure sea levels. Unfortunately, these measurements confirm that sea levels are indeed rising!

5.6 - Ocean emergency: plastics pollution

While walking along the seaside in Nice, I was struck by signs printed on the sidewalk: *Ne rien jeter, tout part à la mer* (Throw nothing away, everything goes into the sea).

Just remembering that we consist mostly of water, should be enough to make us respect and care for water above all else. Unfortunately, this is not the case. Most of what we discard winds up in the sea, and, because much of it is not bio-degradable, we have reached the stage that our oceans, rivers, and lakes are becoming enormous lifeless trash heaps. The oceans are in a state of critical emergency: on the one hand, mankind indiscriminately exploits their resources, on the other, it poisons them by discharging pesticides, petroleum products, industrial and urban waste, and other pollutants. One of the worst pollutants is plastic waste.Plastic materials (polymers), mostly derived from petroleum, were discovered in the middle of the nineteenth century; but it is only since the beginning of the twentieth century that they have become commonly used, if not indispensable, in all phases of our lives. The principal characteristics that render plastics advantageous to other materials are their ease of workmanship, affordability, flexibility, impermeability. They cannot be disposed of by burning, because as they burn they release toxic gases, such as dioxins. Almost all plastic materials float on water, and are not bio-degradable – that is, they cannot be transformed into less polluting chemical compounds by natural agents such as fungi and other microorganisms.

In the ocean, with the passage of time, many of these materials are reduced and eventually miniaturized into miniscule plastic balls, called microplastics, which, when ingested by fish, eventually wind up on our dinner tables. We must therefore ask ourselves how we can best dispose of plastic. There are two ways: one difficult, the other easy. The difficult way is to use no plastic products, or very few; and of the ones we use they must be recyclable or biodegradable. The easy way, the one still most commonly used, is to simply throw them out after use; but - since everything "goes into the sea" – well you see how that goes!

It is estimated that each year between five and thirteen tons of plastic wind up in the ocean. Marine currents deposit this rubbish onto five big underwater islands: two in the Pacific, two in the Atlantic, and one in the Indian Ocean. [Fig. 5.15] The largest of these "garbage dumps"

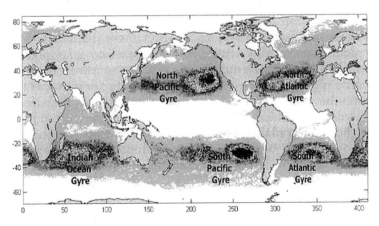

Figure 5.15: *Source, NASA. Throughout the world's oceans there are five zones (gyres) where the garbage of all the inhabitants of our planet gathers.*

– three times the size of France - is in the North Pacific, between the Hawaiian islands and California. We need to also add a sixth island, located in the Mediterranean Sea. The Mediterranean comprises 1% of the surface water of our planet but contains 7% of all discarded plastic. If this maniacal policy of "use and discard" continues, it is estimated that, by 2050, there will be more pieces of plastic in the oceans than fish.

As we've said, we have reached the point where microplastics are part of the food supply of fish, and consequently part of our food supply as well. The solution to the problem lies in the implementation of strict rules to govern plastic production and disposal. All nations will need to cooperate in the effort to stop dumping plastic products in our oceans, so that we can begin to clean them. Currently there are a number of volunteer driven organizations, many organized by young people, which have launched interesting pilot programs to remove floating plastic islands; but the surface areas which must be "swept" are huge and, without a global strategy, cleaning our oceans will be very very difficult.

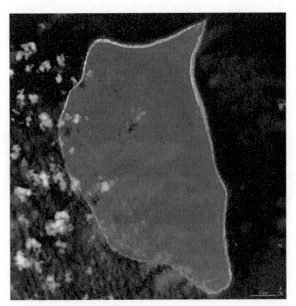

Figure 5.16 *Source, NASA-GSFC. This image taken by the ASTER instrument aboard the Terra satellite on April 3, 2009, shows Henderson Island in the middle of the Pacific Ocean. Although this lush island is located far from any type of "civilization" and is almost completely uninhabited, it is polluted by the most amount of plastic trash on Earth, all of which washes up on its shores – see the white outline all around the island. In order to try to protect it, UNESCO declared it a World Heritage Site in 1988.*

6

SOIL FOR LIFE

6.1 - Chaos, Gaia and Eros

From this height everything was spectacular. But it was the sight of land down there that made me feel secure. I couldn't get over how beautiful everything was. There were so many colors, and so many shades of every color. I began to realize that down there, that very minute, on such a little piece of land, millions of people were working, eating, fighting, or whatever... and I started to really miss it all, and I wanted to get back down as soon as I could. It really is true that Earth is the big Mother - especially since my mother lives there. In one of her few weak moments Caro confessed: I miss my mom, and my dad, and my brother, and my sister, and Andrea, and Antonia, and Camilla, and Fabiana, and Francesca, and... and...and....and after another ten minutes of names I got aggravated and took off. You just can't get too poetic with her.

The mountains! They were full of snow and really beautiful; and the volcanoes! From some of them you could even see plumes of smoke coming out. The giant forests and the deserts! What amazing color contrasts. Some huge clouds of smoke caught our attention, and the Doctor told us they were giant fires that were destroying millions of acres of forest land. "Remember," he said, " almost all those fires were purposely started by people in order to destroy vegetation. It takes about 30 years for a tree to grow, but just 5 minutes to turn it into charcoal."

The rivers were like beautiful ribbons, and when they reached the sea they poured amazing shades of color into it.

And seeing all those lights at night was a spectacular show. When I say night I'm really talking about night down there; up here we'd lost all sense of time because every ninety minutes or so we finished another orbit around the Earth.

We wanted to go home but we had to wait, even though, to be honest, everybody wanted us out of their hair.

We really missed being on solid ground, and sleeping in our own beds, and even though I don't like spaghetti I couldn't wait to get home and eat a monster plate full. We also missed the noise and the confusion.

But if you really think about it, after what we'd done, they probably weren't waiting for us down there with open arms. But I still couldn't wait to get home.

To change the subject the doctor told us the story of Gaia - another name for the Earth.

The ancient Greeks, those know-it-alls, had some very clear ideas about the origin of the Universe. They thought that it was essentially made up of the Earth and the Sky which surrounds it. They believed that in the beginning the only thing that existed was an empty big black hole called Chaos. At some point, no one knows how, out of Chaos came forth Gaia -Mother Earth - which is the opposite of Chaos. Gaia has form. Gaia is order. Gaia is the solid ground on which gods, men, and animals can walk safely. After Gaia came Eros - primordial Love - the Energy that nourishes the universe. At this point, partly because of Gaia, the story gets a little confused. In fact Gaia now gives birth, we don't know how, to two very important children: Uranus, the Sky, and Pontus, the Sea.

Gaia then has a long love affair with Uranus and gives birth to many other children who, unfortunately, couldn't come out of her womb because Uranus attached himself to her and locked them in there. This situation continued until the birth of Cronus (Time), who tricked Uranus into separating himself from Gaia. That's how the Sky finally separated itself from the Earth and the story of the Universe could begin. Her countless children - Titans, Cyclopes, monsters with a hundred arms - were born and consecrated Gaia as the mother of all things: the forests, the mountains, the seas, the skies.

6.2 - The general properties of the Earth

The Earth is a small planet in the solar system. It's more or less spherical in shape with a mass of about 6×10^{24} kg (6 followed by 24 zeroes), an average radius of about 6400 km (3,980 mi), and an average density of about 5.5gr/cm^3.

Although water covers more than 70% of the Earth's surface, its average depth is only about 4 kilometers. Therefore it makes up a very small part of the total mass of the planet. It is estimated that the mass of all the water on Earth is about 1.5×10^{21} kg, or about 4000 times less than Earth's total mass. The atmosphere's total mass is even smaller. It equals approximately 5×10^{18} kg, or 1,000,000 times less than the mass of the whole Earth.

General Characteristics of Planet Earth	
Equatorial Radius (km)	6,378.2
Polar Radius (km)	6,356.8
Difference between Equatorial and Polar Radius (km)	21.4
Circumference of the Equator (km)	40,076.6
Total Earth Surface (km²)	510,000,000
Total Ocean Surface (km²)	361,600,000
Total Land Surface (km²)	148,400,000
Volume of the Earth (km³)	1,083,319,780,000
Mass of the Earth (km)	5.976×10^{24}
Average Earth Density (g/cm³)	5.5

Table 6.1: *General Characteristics of Planet Earth.*

Although the Earth is a small planet, the force with which it pulls everything found on its surface - gravitational force - is strong enough to prevent ocean and atmospheric water from escaping into space.[33]

Temperature and Thermal properties

We've already seen how the surface temperature of the Earth is the result of the combined effects of the sun, the atmosphere, water, and soil. The average temperature of the Earth is approximately 27°C, with variations ranging from about -70°C at the poles to about 60°C in some desert areas.

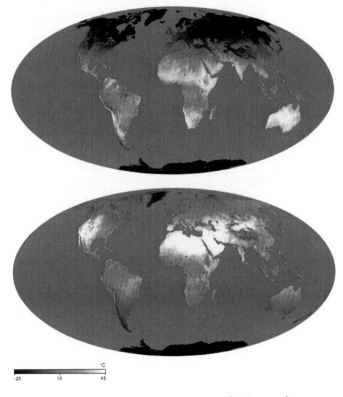

Figure 6.1 *Source, NASA- GSFC, MODIS. Average daytime surface temperatures during the months of January (above) and July (below) in 2003. 2003 was one of the hottest years in Europe. See the temperature scale below the images.*

The thermal properties of soil are very different from those of water. Its specific heat, although different depending on location, is lower than that of water. Its capacity to transmit heat, is generally very high, as is its ability to heat up and cool down.

Eskimos have always been aware of the difference between the thermal properties of water and soil. In fact they build their igloos on frozen waters rather than on land because the ground below frozen ground can reach temperatures less than -40°C, whereas water under ice has a temperature of only about 0°C.

Another important characteristic of the soil, which makes it different from water, is that it is a very good reflector of solar radiation, while water is a good absorber.

Vegetation

Vegetation plays a fundamental role for life on Earth. The quantity and quality of life are closely tied to the quantity and quality of existing vegetation.

Plants and trees are the source of countless types of foods, of fibers we use for clothing, of energy giving substances, of curative substances, etc. And thanks to chlorophyll photosynthesis, plants mitigate the effects of excessive greenhouse gases in the atmosphere by absorbing large amounts of Carbon Dioxide and producing Oxygen.

Since the ground, or whatever is on it, usually reflects solar radiation well, space is an excellent place from which to study and monitor vegetation, both regionally and globally. From above it's possible to study the health of forests and other vegetation by measuring the content of Chlorophyll present in leaves and the surface area they cover.

The method is conceptually easy. The Chlorophyll in leaves absorbs visible solar radiation, while the cellular structure of leaves reflects the radiation in the Near Infrared range. An appropriate combination of these measurements

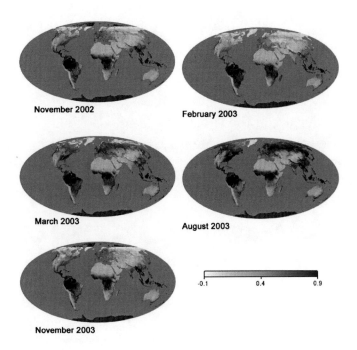

Figure 6.2 *Source, NASA-GSFC. MODIS/Terra. Average global NDVI in different months of 2003. Note the variation (especially in temperate zones) in the NDVI index between Fall-Winter and Spring-Summer.*

allows us to identify the NDVI (Normalized Difference Vegetation Index).[34] NDVI can vary between -1 and +1. Positive values near 1 signify a high concentration of green leaves; values near zero signify a lack of vegetation.[Fig. 6.2]

The vegetation index, being linked to chlorophyll synthesis, is a powerful means of establishing the state of vegetation growth. Periodic observations allow us not only to monitor the state of vegetation but also the effects of droughts and other scarcities.

Composition

We still don't completely understand the composition of the Earth. We know that its average density is about 5.5 times the density of water; and that the density of its surface layer, the Lithosphere (about 100 km thick) is about 2.7 times that of water. What this means is that the volume of a liter of water[35] has a mass of 1 kg, while the mass of the same liter filled with a piece of Lithosphere will be 2.7 kg, and one with "average" earth will have a mass of 5.5 kg.

The density of the superficial layers of the Earth is low compared to the average density of the whole Earth, so that we can conclude that the deeper layers of the Earth must be made up of substances whose density must be about 13 times that of water.

Substance	%
Oxygen	≈ 46.6
Silicone	≈ 27.7
Aluminium	≈ 8
Iron	≈ 5
Calcium	≈ 3.6
Sodium	≈ 2.8
Potassium	≈ 2.6
Magnesium	≈ 2
other substances	≈ 1.7

Table 6.2: *Composition of the Lithosphere.*

6.3 - Characteristics due to Gravity

The Movements of the Earth
The Earth is bound by a periodic movement around the Sun called revolution. A complete revolution takes one year. The Earth's other movements are rotation, on an axis which passes through both poles, and precession of its axis of rotation.

While the Earth revolves around the Sun it also makes a complete rotation around its axis in one "day" or every 24 hours. The most noticeable consequence of this movement is the alternation of day and night. [Fig. 6.3] But this is not the only consequence. As the Earth rotates from west to east every point on its surface also makes a complete turn in one day. This means that points on the equator travel at a much higher speed than those at other latitudes. In Figure 6.4 we can see that point A will move at a different and greater speed than point B. A point on the equator moves at a speed of approximately 1700 km/hr (1,056 mi/hr), while a point placed exactly at either Pole will have zero speed.

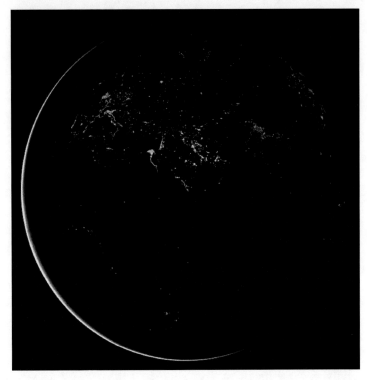

Figure 6.3 *Source, NASA-NOAA. The Living Earth © "Evening arrives over a part of Gaia." This beautiful picture of the Earth is a collage of various images.*

Another effect of rotation is centrifugal force, a force that increases by the square of peripheral speed. This force, because of what we've just said, is greater at the equator than at other latitudes, and being perpendicular to the axis of rotation, it opposes the force of gravity, albeit very weakly.

Another effect of the differences in the peripheral speed of different points on the Earth's surface is the so called Coriolis effect. This is a phenomenon that manifests itself on bodies moving freely (in the absence of friction) from a Pole toward the Equator and vice versa.

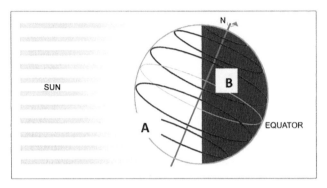

Figure 6.4 *A winter day in the Northern Hemisphere.*

Let us suppose that a body on the Earth's surface is moving freely (without friction) from the North Pole toward the Equator with a constant speed along a certain meridian. If the Earth were not rotating, the body would reach a point on the Equator on the same meridian. However, since the Earth rotates from West to East, the body will be deflected to the right, and will therefore arrive at the Equator at a point to the right of its point of origin, or toward the West. The same thing occurs when a body moves from the Equator toward the North Pole; it will also drift to the right with respect to its direction of movement; that is toward the East.

In the northern hemisphere this effect is responsible for the eastward drift of marine or atmospheric currents (for example westerly winds) which travel from the Equator toward the North Pole; and for the westerly drift of the currents that are moving south (for example the Trade Winds). The exact opposite occurs in the Southern hemisphere.

In the atmosphere, the Coriolis effect is also responsible for the characteristic movement of air currents in Low Pressure zones (Cyclonic) and High Pressure zones (Anticyclonic).

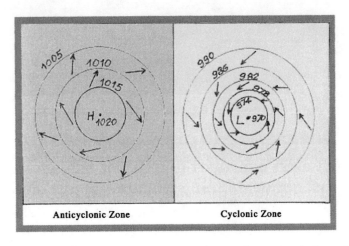

Figure 6.5 *Movement of air currents in the Northern Hemisphere. In an anticyclonic zone, because of high air pressure the air pushes out, and the winds rotate clockwise. In a low-pressure zone air seeks to enter, and the winds rotate counterclockwise. The direction of movement is the opposite in the Southern Hemisphere.*

The formation of Low Pressure and High Pressure zones in the atmosphere is due to the differences in exposure to solar radiation of different places on Earth. At the Equator the air on the surface is warmer than at altitude; as a result ascending currents form, which, like exhaust fans, push air upwards producing areas of low pressure (cyclonic zones). At the Poles the air at the surface is colder than at altitude, so that descending currents form, causing an increase in atmospheric pressure (anticyclonic zones).

Figure 6.6 *Source, ESA. Image from Copernicus-Sentinel 3A, October 11, 2017. Cyclone Ophelia 1330 km from the Azores and 200 km from the coast of Africa.*

Air currents, or winds, move from High Pressure areas to Low Pressure areas in order to create a balance between them. If the Earth did not rotate, these currents would travel radially toward these areas; but due to the Coriolis effect, in the Northern Hemisphere they travel in a counterclockwise spiral. The opposite happens in the Southern hemisphere where they move in a clockwise direction.

The Earth revolves around the sun, together with its satellite, the Moon, in a slightly elliptical orbit whose radius is about 150 million kilometers, and at a speed of about 30 km per second. It completes one revolution in about 365 days, one year. Since the Earth's axis is tilted at a 23° angle to the plane of its orbit, its revolution around the Sun produces the change in seasons, different lengths for night and day, and, as we've seen, varying exposure of its surface to solar radiation. [Shown in Fig. 6.7]

If you've ever played with a top you will have seen that if the axis of rotation of the top forms an angle to the vertical, it will begin a movement of precession around the vertical; in other words the axis of rotation will move along

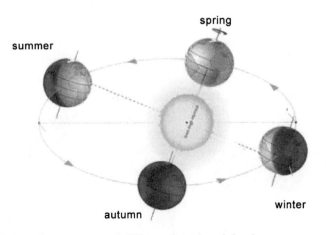

Figure 6.7 *The seasons and different lengths of the day.*

a conical plane. In the case of the Earth the axis of rotation moves around the surface of a cone having an angle at its vertex of 46°. This movement, called Precession, takes about 26,000 years to complete.

The most important consequence of this movement is that in about 13,000 years the seasons will be switched between the two hemispheres, so that in order to find our North we will no longer be able to use the Polaris star, in Ursa Minor, but will have to use the star Vega in the Lyra constellation. [See Fig. 6.8]

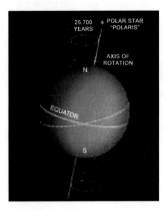

Figure 6.8 *Precession of Earth's axis.*

The shape of the Earth

If the composition of the earth were the same everywhere, and it did not rotate, it would - because of gravity - be a perfect sphere. The only geometric shape that would cause the force of gravity to be the same at every point on Earth, and always directed at its center, is a perfect sphere. A spherical surface is also called "equipotential" because if we measure the weight of a body - really the force with which the Earth attracts it - we would get the same value everywhere on Earth. But this is a very approximate model which does not account for the effect of the Earth's rotation. In fact, because of rotation, points at the Equator travel at a much higher peripheral speed than those at other latitudes.

The result is that centrifugal force will oppose the force of gravity differently at different latitudes, so that the weight of a body will be less at the Equator that at the Poles. In addition centrifugal force tends to push matter near the Equator outward causing a bulging of the sphere with respect to polar regions, deforming it into the shape of a "rotational ellipsoid" whose radius at the equator is about 21 km longer that it is between the poles.

The surface of the Earth is also not smooth, but very "pockmarked." In fact there is a difference of about 20 km between the highest mountain and the lowest depression. In addition the different materials that make up the mantel and crust of the Earth are neither homogeneous nor uniformly disposed. For example, the Earth's crust below the oceans is much finer and denser that the continental crust.

All of these factors mean that the Earth's gravitational pull varies from place to place, and it isn't the same even at those points located at the same distance from its center.

Once again space is an exceptional place to study the variations in the force of gravity. In fact since the orbit of an earth satellite depends on the value of the gravitational pull of the place on Earth above which it finds itself, its elliptical orbit will not be on a constant or fixed plane but will slightly deviate from it. In 1959 the study of the deviations from its elliptical orbit of the American Vanguard satellite allowed us for the first time to determine with great precision the form of Earth's Geoid; that hypothetical equipotential surface which would cause a marble placed upon it to remain still, just as it does when we place it on a horizontal plane.

Since water always assumes a surface perpendicular to the force of gravity, and a plumb line is always parallel to the force of gravity, a Geoid is a surface which will always be perpendicular to a plumb line. The Geoid then is that hypothetical surface which is the same as an ocean at rest, without tides or waves. In the open ocean far from coasts, the Geoid coincides with the mean ocean surface. If on land there existed an imaginary network of canals connected to the ocean, and in the absence of friction, the Geoid would be the surface level assumed by the canals.

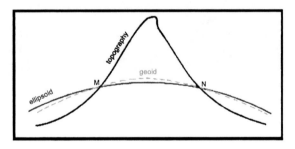

Figure 6.9 *Comparison of the physical surface (topographical), the ellipsoid, and the geoid.*

A Geoid surface, in other words, is the shape the Earth would have if completely covered by water. The Geoid doesn't coincide with the rotational ellipsoid because the interior of the Earth contains different masses, and because its surface landmass is irregular. The Geoid is higher than the ellipsoid with respect to the surface landmass, but lower than the ellipsoid with respect to the oceans. [Fig. 6.9] The surface of the Geoid can differ from the ellipsoid even by tens of meters.

Figure 6.10 illustrates the surface of the Geoid as reconstructed with measurements taken by the ESA GOCE mission in 2011. The red areas show the Geoid higher than the ellipsoid by as much as 85 meters, while the blue areas show it lower by as much as 106 meters.

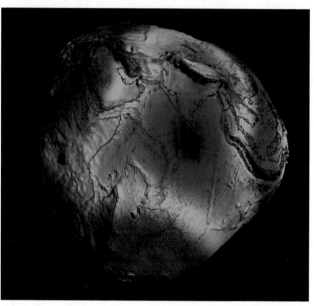

Figure 6.10 *Source, ESA-HPF-DLR. Difference between the Geoid and the ellipsoid as measured by the GOCE mission in 2011. Be careful, this is not the true shape of the Earth.*

The surface of the Geoid is not constant or smooth, but has continuous slight variations. In fact, since gravity is tied to mass, every movement of matter - be it instantaneous as in earthquakes, or slow as in ocean currents or melting glaciers - will cause a change in the value of the gravitational field at that location; and therefore a change in the Geoid. It has been calculated that as a result of the terrible Indian Ocean earthquake of December 26, 2004 - magnitude 9 on the Richter scale - the Geoid has shifted by as much as 18mm.

A precise knowledge of variations in the Geoid gives us a powerful tool to understand important phenomena tied to global climate changes; to better understand the distribution of masses in the Earth's interior; the shifts in its crust due to earthquakes; the places where volcanic lava may escape; the location of oil fields, of mines, etc.

Finally, since the Geoid constitutes the natural "horizontal" plane of the Earth, it is utilized as a surface of global reference from which to map all the topographical characteristics of the Earth, be they the landmass, the oceans, or glaciers.

6.4 - Soil emergencies

The solidity of the ground gives us a sense of stability and safety which is absolutely false. In fact the Earth's crust is in continuous evolution. Evolution caused not only by what happens in the deepest interior of the Earth (for example volcanic eruptions, seismic events, etc.), but also by surface phenomena (changes in coastlines due to changes in sea levels, changes in desert formations due to winds, changes due to landslides, fires, etc), and finally by our own handiwork, in other words by the uses we make of the soil (agriculture, urbanization, roads, dams, artificial lakes, etc.).

Human and biophysical factors such as the need to meet ever increasing demand for food, energy, potable water, and urban expansion, bring about significant changes in soil cover and usage. These changes manifest themselves in various forms: expansion of urban areas, expansion or abandonment of agricultural areas, deforestation or reforestation, construction of dams, fires, floods, droughts, insect infestations, etc. Some of these events cause changes which can range from a day or so, to a few weeks (storms, floods), to a few years (changes in agricultural practices), to many years (deforestation, reforestation, the effects of fires, etc.).

In order to understand how the Earth-system reacts to natural (out of our control) or man made changes, and to be able to predict the effect these changes may have on climate, or their connection to natural catastrophes, it is particularly important to understand the changes in the uses and covering of the soil.

Often in common parlance we refer to certain catastrophes as "natural," when we should more accurately call them "announced" catastrophes, since they occur because of neglect or poor use of the Earth-system, and should, with common sense, be foreseeable. Some of these catastrophes have reached such enormous proportions - in lives affected, areas stricken, and duration - that the United Nations has defined them soil emergencies.

Deforestation

Forests have a fundamental role for life on Earth. They are immense furnaces in which countless forms of life (including humans) not only find a home, but they are continuously subjected to changes and evolutions. Forests act as large absorbers of Carbon Dioxide and at the same time large producers of Oxygen. They exchange energy and humidity with the atmosphere. They control soil erosion and the course of surface water.

Currently forests cover about 30% of the Earth's landmass, and are principally located in the Amazon, Canada, Russia, Africa, and Southeast Asia. The Amazon rainforest is the largest forest on Earth. Although it occupies about 5% of the Earth's landmass, it is capable of capturing about 10% of the world's Carbon Dioxide and "warehousing" it in biomass such as branches and leaves.

Unfortunately population increases, poverty, the exploitation of the soil for housing and road construction, infrastructure, and an increase in the production of agricultural products, has caused, especially in the last century, a destruction of forested lands. Currently we are destroying, every year, an area of forest equal to about half the size of Italy; and this statistic will worsen in the coming years. This fatal process is called Deforestation.

Deforestation carries grave consequences both for climate and the soil. A lack of trees means less water evaporation from plants, an increase in greenhouse gases, and therefore an increase in local temperature. Moreover, less plant life means that rains will cause more soil erosion.

In the Amazon Rainforest these events, together with habitat destruction, have killed many forms of life in the local ecosystem. [Fig. 6.11]

Other factors which cause deforestation, besides tree cutting, are plant diseases, soil and air pollution, and fires. [Fig. 6.12]

Forest fires sometimes start naturally, but are often caused by humans. In many tropical regions - where forest concentration is higher - it's common, because of terrible poverty and population growth, for people to start forest fires in order to obtain more arable land. Unfortunately the resulting fields are difficult to cultivate because the soil beneath the forest is very poor, and they are quickly abandoned. Reforestation, even if desired, would then take about 50 years.

Fires always produce nefarious environmental consequences. In addition to the effects we've mentioned (decreased evaporation and emission of plants' stored Carbon into the atmosphere), they also eject large quantities of aerosols into the atmosphere.

As we can see, the problem of deforestation is

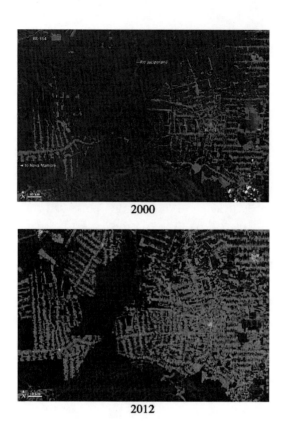

2000

2012

Figure 6.11 *The state of Rondônia in the western part of Brazil. An area of the rainforest of 208,000 square kilometers has been rendered one of the most deforested areas between 2000 and 2012. Source, NASA-GSFC. MODIS Terra*

due to strong societal motivations, and its solution must be based on a global commitment of all of mankind. The United Nations has declared deforestation to be one of the world's principal environmental emergencies.

Desertification

Another grave emergency for humanity is desertification. By desertification we mean a degradation of the soil's capacity to sustain the biological functions of animal and vegetable life. The causes of desertification are both

Figure 6.12 *Source, NASA-GSFC, MODIS Terra and Acqua. Active fires (in red) in the Brazilian rainforest during the week of August 15-22, 2019.*

natural and man made. The process occurs in different geographical regions, and is extensive; about one third of the Earth's landmass is threatened by it, and it is closely linked to deforestation. Desert areas are advancing inexorably, and without prompt and methodical actions by mankind they will continue to do so. Data obtained from space has shown that in the last ten years the Sahara, the largest desert on Earth, has been in constant expansion. [Fig. 6.13]

Certain regions of southern Europe are already affected by phenomena typical of desertification.

When the surface of the Earth is struck by solar radiation it reflects the visible part back into space and absorbs the infrared part, causing the humidity present in the soil to evaporate. If the soil lacks the elements capable of retaining or refurnishing it of humidity, the Earth's

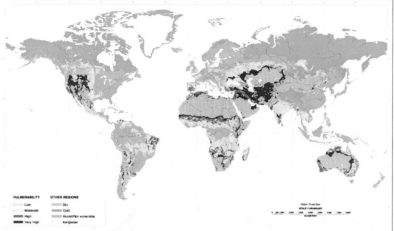

Figure 6.13 *Source, USDA-NRCS, Soil Science Division, World Soil Resources, Washington D.C. FAO. The four zones of desertification vulnerability based on climate and soil classification. Note that the zones at risk of desertification are 43% of sea level areas which contain 35% of the world's population.*

surface will become more and more arid since the humidity present below the surface will rise to the surface and be constantly evaporated. The result will be a dry and pulverized surface: a desert. The climate in desert zones is very hot during the day and very cold at night.

Desert advance is occurring in vast areas of Africa, Asia, Latin America, and Australia, affecting an estimated 250 million people, and causing extremely severe human and economic losses. What can be done to stop this phenomenon?

Certainly the causes are many and very complex. But, as with deforestation, human activity is an important cause of the increase in desertification. Our understanding and commitment are the only instruments we have at our disposal to solve this problem.

Once again space is an advantageous place from which to understand the causes of desertification, and how

140

it progresses. Taking into consideration the increasing disappearance of forest areas, the increase in desert areas, the ever increasing population of the world with its concomitant human needs, and that the pro capita "productivity" of the planet is greatly diminishing, we are faced with a terrible emergency in terms of the future of humanity, which only a global effort can resolve.

7

THE CYCLES

7.1 - The Return

The six days of our stay in space had gone by in an instant; we'd orbited the Earth dozens of times, and now we had to get ready to come back.

Commander Gaiova told us that our little escapade was almost over. She also said that reentry was going to be less traumatic than take-off, but that there were other risks involved. The Doctor had explained to us that when the shuttle - going about 20,000 kilometers an hour- came into contact with the first layers of the atmosphere some of its parts, because of friction with the air, would reach temperatures in the thousands of degrees, temperatures that few materials could resist.

Being sensitive to motherly instincts the Commander had organized a video conference with our parents to calm them, and us, down a little.

They all told us there was nothing to worry about because everything was going to go smooth as silk, and they couldn't wait to hug us again. But since whenever we went anywhere they were always harping on us to be careful, watch out for cars, don't talk to strangers, etc., this advice now instead of calming us down made us more nervous.

The crew for the return trip was different from when we took off. It was made up of Commander Gaiova, the co-pilot Paul Cadorinne, us two, and three astronauts who'd been on the Space Station for some time. Two of these guys were Russian, and the other one was a filthy rich American tourist who, they claimed, had paid about 50 million bucks to spend a month in space. And we hadn't spent a dime! That's what I was going to tell my father; that by stowing away I'd saved him so much money that even if he lived a thousand years he would never be able to afford such a trip!

From the moment we undocked from the Space Station until we landed on Earth very little time went by, maybe an hour or two. Little by little as we got closer to Earth we felt a weird sensation all over; after a while we started to realize what it was: it was our weight. It was like the feeling you get when you get off a roller coaster.

The critical time when the shuttle entered the first layers of the atmosphere at incredible speeds passed by without any problems. The shuttle's thermal protections and the Commander's skills both did their job.

The landing was almost like a regular passenger jet. A couple of scary minutes, a bumpy touch down, a jamming on the brakes that brings your stomach into your mouth, and then a big sigh of relief.

Before we left the ISS I got a little chocked up when we said good-bye to the Doctor. I knew he was happy that we were getting out of his hair and he was finally free to get back to his scientific experiments. But I was sure that during all the time that he'd been babysitting us he got attached to us and to all the silly things we said and did. Speaking of silliness, Caro couldn't pass up one more chance to get stupid. When Doc was saying good-bye to us he was pretty lovey dovey, at least as much as a guy can be who's got no social contacts whatsoever; a lonely bachelor who talked with a thick german accent. (Shows you what I know, because when he came back to Earth this beautiful woman was there to greet him! Boy do I have a lot to learn!).

He reassured us about the reentry and told us not to worry about the chicks, that they would be born in a couple of days. Caro asked him if it was true that when an animal is born it recognizes the first living thing it sees as its mother.

Doc answered that there was some truth to that, and started talking about "imprinting", about this guy Lorentz and his geese, etc. Caroline gave him a big hug and as she was saying good-bye she told him that you couldn't pay her enough to miss the chance to see him walking out of the shuttle surrounded, like a good mother hen, by about twenty chirping little chicks. Wouldn't miss that show for anything!

When I saw Doc looking at her with this serious expression on his face, I realized he hadn't thought about that possibility. I gave him a big warm hug too, and I thanked him for all the time he spent with us. He gave me a little peck on the cheek and ran off to all his projects.

7.2 - The Steam Engine and the Earth-system

Few of us have ever seen a working steam powered train, although we've probably heard about them and their choo choo sound; their heart is the steam engine.

The steam engine was the fundamental invention that propelled mankind into the industrial age. The steam engine is a thermal machine, a mechanism which, by exploiting the properties of fluids, usually water, can transform heat (thermal energy) into mechanical energy.

The mechanism of this machine is conceptually very simple. In a container, called a boiler, a certain amount of water is brought to boil and evaporates. The vapor is forced through a cylinder and pushes a piston; the piston, by means of a connecting rod, causes a crank shaft to turn, which then turns a wheel. [Fig. 7.1] When the wheel completes its turn it brings the piston back; the piston now pushes the water vapor into a coil where it condenses and becomes water again, which goes back into the boiler to begin the process all over again. We call this process (a fluid undergoing this type of transformation and returning to its original physical state, like water in a steam engine) a thermodynamic cycle.

There are many different kinds of thermal machines depending on the types of fluids and the types of cycles to which they are subjected; their operation is governed by the Second Principle of Thermodynamics. This law affirms that no machine exists that can transform all the heat it receives from a source at a certain temperature into mechanical energy; but that it will, by necessity, give up some of the heat to a source at a lower temperature. This is exactly what happens in our steam powered train. During its cycle the steam engine receives heat from the boiler, transforms part of it into mechanical energy, and then releases part of the heat in the condenser which is at a lower temperature than the boiler.

Figure 7.1 *The Rocket: One of Robert Stephenson's first steam engines, 1829.*

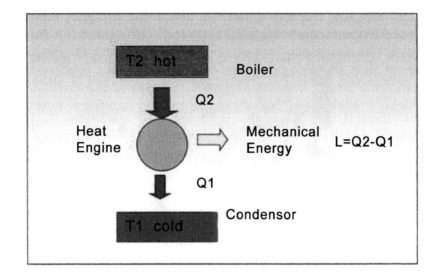

Figure 7.2 *Thermal engine. The engine absorbs Heat (Q2) from the Heat Source (T2), gives up a portion (Q1) to the Cold Source (T1) and produces a certain amount of labor (L): L = Q2 - Q1.*

The Earth-system is also subject to this Principle. The two fluids, water and atmosphere, subjected to cyclical transformations, transfer thermal energy between the Earth's surface (the Hot Source) which has a median temperature of 15°C, and a Cold Source - the atmosphere at an altitude of about 5 km, which has a temperature of about -18°C. The mechanical energy or Work produced by this machine moves the same fluids and is then dissipated in attrition. In the steam powered train the energy needed to feed the engine and start the Cycle was furnished by the combustion of coal that a couple of poor machinists shoveled into the furnace under the boiler. Needless to say in the case of the Earth-system the source of the needed energy is the Sun.

7.3 - The Hydrologic (Water) Cycle

We believe that water appeared on Earth about 4 billion years ago. It may have been produced by volcanic eruptions or by comets or asteroids which hit the planet, we don't know for sure how or why, but luckily it happened.

The quantity of water present in the Earth-system, in its three states, solid, liquid, and vapor, is constant and passes continuously from one state to another. This is called the Water Cycle: a continuous passage of water molecules from the surface, the Lithosphere and Hydrosphere, to the Atmosphere, and vice versa. [Fig. 7.3]

What provides the energy which fuels this Cycle? The Sun naturally!

The liquid water present on the Earth's surface is constantly evaporating, ie becoming gas. It is estimated that this process accounts for about 90% of all the water present in the atmosphere. The other 10% is delivered by plant transpiration, the process by which plants absorb water from the ground and carry it up to their leaves, where it is then released through the stomata, tiny pores on the underside of leaves. Other processes which contribute to the transport of water vapor to the atmosphere, albeit in very small amounts, are sublimation, or the transformation of ice into vapor, and the water vapor emitted by volcanic eruptions.

As water vapor rises into ever colder heights in the atmosphere, the air cannot hold on to all of it. The excess amount of vapor condenses and returns to earth in the form of rain, snow, or hail.

We have already seen how this cycle allows for a transfer of energy from one part of the system to another.[36] The amount of water that this cycle moves is really enormous. In fact we calculate that the quantity of water which, in one year, evaporates, condenses, and returns in precipitation equals about 500,000 billion tons. Whereas the amount of water constantly present in the atmosphere is only about 13,000 billion tons.

It may seem strange that although such a great quantity of water circulates in the atmosphere, a very small amount remains there permanently. To better understand this oddity let's think about a bank teller who only has a hundred dollars in her purse, but who is charged every day with counting all the money deposited in the safe. A great amount of money will probably pass through her hands on a daily basis, but when she goes home at night, if she's honest she'll still have only a hundred dollars.

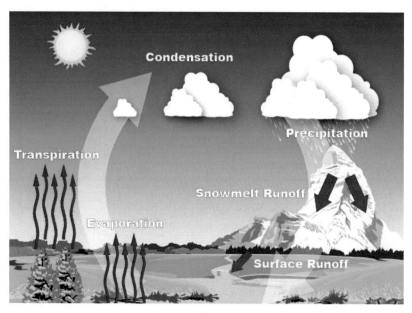

Figure 7.3 *Source, USGS http://ga.water.usgs.gov/edu/watercycle.html Illustration: John Evans: The Water Cycle.*

The water cycle is one of the two engines (the other being the Carbon cycle) of the climatic system: from floods to droughts, from irrigation to the water we drink every day, from meteorologic events to the state of forests, from life to death, everything is tied to water. [Figure 7.4]

Is human activity changing this cycle in any way? Since the total amount of water in the Earth-system is constant, changes in the cycle can only be those that move more water from one phase to another. If, for example, the temperature of the oceans increases, evaporation will also increase. An increase in the amount of water vapor in the atmosphere, ie an increase in the amount of water present, will bring about an increase in precipitation. In addition, since water vapor acts as a greenhouse gas, the Earth's temperature will also increase, which in turn means more evaporation. From this simple scenario we can conclude that if human activity causes the temperature of the Earth to go up, the hydrologic cycle will change as a consequence. We can see then, how important it is to study this cycle and all the phenomena that can, even minimally, "force" it to undergo any modifications.

7.4 - The Carbon Cycle

After Hydrogen, Helium, and Oxygen, Carbon is one of the most abundant elements in the Universe and one of the most abundant on Earth. But its presence is even more important for us because Carbon is the fundamental element needed for building all the organic materials indispensable for almost all forms of life.

Some theories maintain that Carbon has been present on Earth since its birth, that is since 4.5 billion years ago.

Carbon is found in the soil and on the soil, in the oceans, and in various chemical forms in the atmosphere: pure Carbon, Carbonates (Calcium, Magnesium, etc.), Carbon Dioxide (CO_2), Carbon Monoxide (CO), etc.

It is estimated that the Carbon contained in the soil

amounts to about 1,580 GtC,[37] in vegetation about 680 GtC, in the oceans about 38,100 GtC, and in the atmosphere in the form of CO_2 about 750 GtC. [Fig. 7.4]

Part of the Carbon in its various chemical forms is exchanged cyclically between the soil and the atmosphere, the ocean floor and the earth's interior, the soil and the seas, the oceans and the atmosphere.

There are two types of Carbon cycles: one which takes hundreds of millions of years, known as the Geologic Carbon Cycle, and the other which occurs almost yearly, known as the Biological Carbon Cycle.

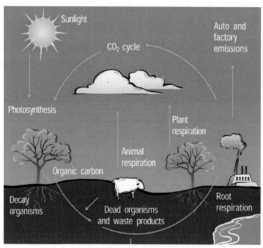

Figure 7.4 *Source, U.S. DOE. 2008. Carbon Cycling and Biosequestration: Integrating Biology and Climate Through Systems Science; Report from the March 2008 Workshop, DOE/SC-108, U.S. Department of Energy Office of Science (genomicscience.energy.gov/carboncycle/).*

The Geologic Carbon Cycle

The Earth's crust is made up in large part of Carbonates. Rain is responsible for the formation of Carbonates by carrying the Carbon contained in the atmosphere as Carbon Dioxide back to Earth, and allowing it to combine with its respective minerals. Then, by the process of erosion, rains carry Carbonates into the Oceans where they are deposited on the ocean floor forming layers of sediment. The slow movement of tectonic plates drops Carbonates into the Earth's interior, where, due to extremely high temperatures, they fuse and release their Carbon. Volcanos then finally return the Carbon to the atmosphere, thus fueling the cycle again. [Fig. 7.5]

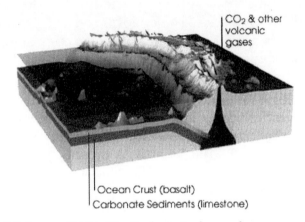

CO$_2$ & other volcanic gases

Ocean Crust (basalt)
Carbonate Sediments (limestone)

Figure 7.5 *Source, NASA. The Geologic Carbon cycle.*

The Biological Carbon Cycle

The biological cycle is an accumulation of processes by which great quantities of Carbon move from the Oceans to the Atmosphere and from the Soil to the Atmosphere, but in such a way that the net flow in a year is zero. The amount of Carbon moved in this cycle is approximately one thousand times greater than in the geologic cycle.

The major Carbon exchange between the soil and the atmosphere occurs in two fundamental processes: Photosynthesis, or the production of Glucose (a sugar), and Respiration, the opposite of Photosynthesis, or the transformation of Glucose into the energy needed by plants to grow and reproduce. In Photosynthesis, Glucose is produced by using solar energy to remove Carbon and

Water from the atmosphere (present as Carbon Dioxide and Water Vapor). In Respiration the energy absorbed during Photosynthesis is returned, and Carbon is released into the atmosphere as Carbon Dioxide. In the light of day the effects of Photosynthesis are much more important than those of Respiration. But in time, the process of Respiration together with the Decomposition of organic matter, brought about by bacteria and fungi, restores all the removed Carbon back to the atmosphere. The processes of Photosynthesis and Respiration-Decomposition slow down when atmospheric temperature is too cold or the climate is too dry. This is why vegetation greatly diminishes in winter months, especially in temperate zones.

As we've said earlier, the large Carbon Flux (about 120 GtC) that Photosynthesis and then Respiration cause to circulate between the soil and the atmosphere is subject each year to fluctuations in the amount of atmospheric CO_2; but its average yearly value is zero.

The biological cycle also plays an important role in the geologic cycle. For example Phytoplankton use the Carbon and Calcium in the ocean to build their microscopic shells. When the Phytoplankton die these shells collect on the marine floor and take part, as we've seen, in the geologic cycle. In the same way Coal and Petroleum deposits, of organic origin, constitute reserves of Carbon on a geologic scale.

Fires play a fundamental role in the transfer of Carbon Dioxide from the soil to the atmosphere. And deforestation also causes an increase in the amount of CO_2 the Earth exchanges with the Atmosphere. It's important to note that a dead tree is almost 50% Carbon. Carbon thus produced is warehoused in the Atmosphere. Only the spontaneous rebirth of vegetation in burned out areas, or reforestation, can re absorb it, thereby decreasing the amount of Carbon Dioxide in the Atmosphere.

In the Oceans the absorption of CO_2 from the Atmosphere is controlled by the temperature of the water and the quantity of CO_2 it contains. In the oceans, like on land, Photosynthesis and Respiration of plants and marine microorganisms are the two processes that absorb and release Carbon Dioxide. But unlike in the Atmosphere-Land cycle, the amount of Carbon Dioxide exchanged between the Atmosphere and the Oceans (about 90 GtC) occurs in a cycle of very short duration, from a few days to a few weeks. And an even more important difference is that the Oceans do not accumulate Carbon deposits.

Is human activity changing in some way this natural Carbon exchange among Earth, Atmosphere, and Oceans? Unfortunately yes!

We've already seen how deforestation and fires contribute to an increase in the amount of Carbon the soil exchanges with the Atmosphere. To this we must now add the contribution of human activity. A more intensive use of our soil, together with the ever increasing use of fossil fuels (used by industry, in energy production, by automobiles, etc.), have become the source of a consistent increase in the quantity of Carbon that, as CO_2, is deposited in the Atmosphere.

Carbon Dioxide levels measured in the last fifty years or so, in places far removed from human activity, indicate a marked increase in CO_2 concentration during this period. [Fig. 7.6]

Currently we are seeing a yearly increase in CO_2 in the Atmosphere of about 7.1 GtC . Industrial activity accounts for 5.5 GtC and deforestation for 1.6 GtC. Not all of this enormous quantity of Carbon Dioxide remains in the Atmosphere. We estimate that about 5.2 GtC is recycled between Atmosphere-Earth and between Atmosphere-Oceans, while the remaining 1.9 GtC is the yearly contribution of human activity to the 750 GtC warehoused in the Atmosphere. If things remain as they are, mankind will cause a .2% yearly increase in the amount of Carbon in the Atmosphere. This means that if things remain as they are in less than ten years the increase will be 2%, and in less than one hundred years 20%! This increase will cause a concomitant increase in the Greenhouse Effect!

It bears reflecting that even if all the Powers on Earth decided today with a magic wand to stop this constant increase, it would still take many years to see appreciable changes.

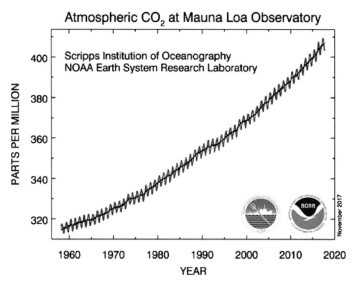

Figure 7.6 *Source, NOAA (R. Simmons). The CO_2 concentration in the atmosphere has increased in the last 50 years from 310 to 360 ppm (parts per million).*

8

THE BIOSPHERE

8.1 – Easter Island

In one of the most remote regions on Earth, about 3,800 km from the coast of Chile, there is an island called Rapa Nui,[38] or Easter Island. The ancestors of its current inhabitants believed that the first people who populated the island arrived there, around the sixth century AD, in two catamarans piloted by a brother and sister; they labelled it "the navel of the world."

The first western people who set foot on the island found it almost totally uninhabited and devoid of any stands of timber. However they found ruins[39] attesting to the existence of a previous substantial population living among vast forests of giant palms. In contrast to its present misery they saw evidence of a past landscape both rich and mysterious, and discovered about nine hundred megalithic statues, called Moai, placed along the coastline facing the ocean, as if awaiting someone's arrival.

Many questioned why there were so few people on the island. What had happened to the giant palm trees? How had the natives carved and transported the gigantic Moai statues? The exact answers to these question have yet to be found, but the most popular current theory is as follows: ambitious chieftains eager to create the great monoliths, ordered that giant palm trees be cut down, since they alone could furnish trunks thick enough to transport the blocks of stone from which the Moai statues would be carved. The continuous destruction of the forests, coupled with a number of fratricidal wars, eventually brought about the desertification of the island, and the near total disappearance of its inhabitants. (Others believe that there were other or concurrent causes, among them diseases brought over by Europeans).

This story should make us reflect! Rapa Nui, isolated in a vast ocean, with finite yet sufficient resources to maintain a substantial population in relative well- being, in many ways resembles our own planet. The Earth too is isolated in space, is rich in great yet finite resources, is often governed by ambitious and greedy rulers, and is inhabited by many selfish individuals who place their own wellbeing above the common good, and who are ready, like those islanders, to cut down "the last palm" in order to erect their own Moai.

8.2 – The Biosphere

The biosphere is the Earth's skin: a very thin layer which extends about 10 km above and 10 km below sea level. In it a most incredible and mysterious phenomenon occurs: Life! Everything which is alive, or which permits life, occurs in the biosphere: plants, seas, air, fish, birds, insects, bacteria, etc. Within it, solar energy, air, water, and soil continuously interact with each other exchanging "energy" in order to foster life.

Throughout the long history of our planet great geological and climatic events have disturbed and modified the biosphere, and consequently all life forms on earth. One hundred million years ago the absolute rulers of the biosphere were the dinosaurs, but for the last one hundred thousand years or so human beings have become its masters.

In the last two hundred years - a very brief period of time with respect to natural evolution, which has occurred

156

on our planet for four and a half billion years - humans have reached a state of development capable of profoundly modifying and utilizing the biosphere for their own needs. This has been due to the difference between human intelligence and the intelligence of other animal species. Because humans are the primary rulers, and therefore the cause and reason for these changes, the current geological epoch is called Anthropocene.

Do we deserve this role? Human changes to the biosphere are essentially caused by population growth and its consequences. There is an ever increasing need for food, energy, land for habitation, transportation, consumer goods, etc. Believing that our planet's resources are unlimited, we have intensified the competition between demographic growth and industrial growth: more people means more industry, more land needed for agriculture, more pesticides, more fertilizers, more air pollution, more acidification of our waters, more non-biodegradable refuse, newer diseases, more destruction of biodiversity.

Have we, like the rulers on Easter Island, committed the sin of greed? For surely if we continue along this path we will inevitably approach the precipice of our self-destruction.

8.3 – Demographic growth

In the last two hundred years or so demographic growth has followed a vertiginous slope. According to United Nations sources, from 1800 to 1927 (127 years) the world's population grew from one billion to about two billion; from 1927 to 1974 (only 47 years) it doubled to four billion; then in just 25 more years it increased again to six billion; as of 2010 it had reached seven billion, and now in 2023 it is about eight billion. We are now increasing at a rate of about 1 billion more people every twelve years or so. As we can see in Fig. 8.1, it is estimated that this rate of growth may begin to decrease around 2050, a year in which the world population may reach nine billion, and that by 2100 it may be about 11 billion.

The graphs in Fig 8.1 reveal, however, that demographic growth is not uniform: it is greater in poorer continents than in richer ones. For example, Africa has the greatest rate of growth: between 1950 and 2050 its population will have gone from 200 million to 2 billion.

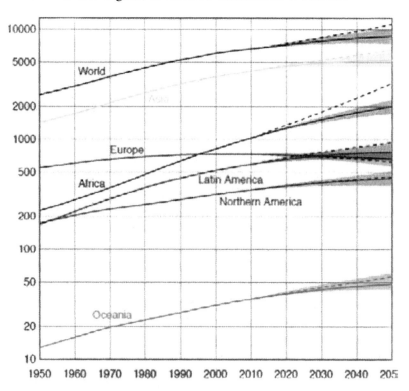

Figure 8.1 *Source, United Nations, Department of Economic and Social Affairs, Population Division (2015). World Population Prospects: the 2015 Revision. Population on different continents and globally from 1950 to 2050. Starting from 2015 only if the estimated values and relative uncertainty appear.*

From ancient times, until the first half of the nineteenth century, human population and its basic necessities generally grew in a balanced manner: that is, nature was able to replenish what humans took. But with the discovery of the "machine" capable of doing work using chemical rather than natural energy, the man-nature balance has been irreparably broken. The benefits attained have been enormous, allowing us to grow and prosper; but the Earth's resources are not limitless, and the laws of nature teach us that there is a price to pay for all the changes (good or bad). Therefore we must prepare ourselves to face our "debts" before we go bankrupt. As early as 1972 a group of scientists, intellectuals, economists, and business leaders known as The Club of Rome, issued a report entitled "The Limits to Growth," in which they warned that uncontrolled population growth would cause drastic changes in the biosphere, putting at risk the very survival of our own animal species.

8.4 – The reduction in biodiversity

The most noticeable and worrisome effect on the biosphere brought about by the combination of industrial development and population growth is, without doubt, the reduction in the diversity in the biological world.

The biodiversity of a region is measured by the number of different species that live in it, from mushrooms to humans, from plankton to sequoias. Obviously there are areas on earth where biodiversity is greater, as in temperate and tropical zones, and areas where it is much lower, as in desert and polar regions.

Biodiversity is similar to a giant puzzle in which all the pieces (species) are interconnected. Each piece plays a role which cannot be played without the help of other pieces. In the earth's biosphere humans are but one piece connected to many others, and whose existence is therefore tied to the existence of every other piece.

We can think about our food chain in the same way. The sun provides the energy for photosynthesis; photosynthesis fuels the growth of vegetation in soil rich in nutrients; animals eat vegetation, and humans eat animals and vegetation; when an animal or a human being dies, it decomposes and promotes the growth of more vegetation.

There are numerous examples of this cycle, but the process is always the same: the sun furnishes the energy which is transformed by each component of the biological worlds so that it can nourish the component above it in the chain. All of biodiversity literally contains "a little piece of the sun."

What would happen if fungi, which decompose matter necessary for the food chain, disappeared? What would happen if insects which pollinate plants disappeared? What would happen if any piece of the puzzle disappeared? The answer is clear: we would need to find a substitute or little by little all the other pieces would become weaker, break down, and the whole puzzle would collapse.

Fortunately the biosphere has not yet reached the point of no return. But it is perilously close. It is estimated that there are circa 8.7 million species on Earth, with a margin of error of 1.3 million. About 1.5 million species have been identified, 17% of the total. There is much to be done in order to discover all the species which exist, but we are already witnessing the extinction or threatened extinction of many species closest to ours. There are many causes of this dangerous phenomenon, but most lead back to one source: human beings.

As the human species increases, many animal and vegetable species are disappearing. Think about the near extermination of the bison in the late 1800's; the indiscriminate hunt for whales by very "civilized" countries like

Japan and Norway; the slaughter of elephants and rhinoceroses for their ivory tusks; the continuing destruction of the Amazon rainforest (the largest lung of our planet) to clear land for agriculture; modern industrial fishing methods which destroy marine habitats; the indiscriminate use of pesticides on a massive scale which kill all types of insects (both bad and good); the migration of invasive alien vegetable and animal species which destroy indigenous ones; the pollution of the oceans; and on and on...

In order to monitor the dangers of species extinction, the International Union for Conservation of Nature (IUCN)[40] was created in 1964. This body, which unites scientists from all over the world, is financed by private donations, industries, and international institutions. The principal activity of the IUCN is the compilation of the so-called Red List, an inventory of the state of health of biological diversity on Earth. [Fig. 8.2] This list, which scientists prefer calling the "Barometer of Life on Earth", reports what actions must be undertaken to save those species at greatest risk of extinction.

The main objective of the Red List is to examine in detail the state of 160,000 species by the year 2020. Scientists have currently analyzed more than 90,000, and the results are worrisome: more than 27% are in danger of extinction. The species most at risk, in descending order are: amphibians, conifers, coral reefs, sharks, ray fish, mammals, and birds.

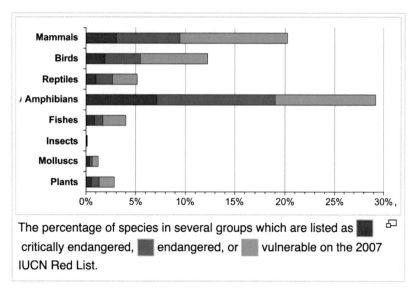

Figure **8.2** *Source, IUCN (2007).*

9

ARISTOPHANES AND GLOBAL WARMING

9.1 - Peace

Well we'd been back from our "getaway" for a pretty long time. We got to be celebrities for a couple of minutes, but then back to the usual stuff, like school six days a week. After a couple of days of partying and a lot of questions, our friends brought us back down to earth after they realized we hadn't changed a bit, that we were just like them...and that everything that happened to us was because of a bunch of lucky accidents. Truth be told they didn't put it so politely, if you know what I mean.

We went with our literature teacher to see this play Peace by a Greek writer called Aristophanes. We were happy to go to the theater because it meant getting out of school for five hours; but when we found out that the play was by this ancient writer, and with such a lame title, we planned on taking a nice long nap; unless of course something more exciting turned up at the theater.

But once the play got going, and I got into it, I changed my mind. Even though this comedy was written hundreds of years ago it was pretty interesting and really up to date, and unbelievably ironic.

The plot was all based on the story of the Peace Goddess and the War Goddess.

A long time ago the Goddess of Peace ruled the Earth and everything was cool; the farmers grew a lot of food, and the people were happy. But all this good feeling unfortunately didn't last long, because one day the Goddess of War, with a trick ,captured the Goddess of Peace. She locked her up in a cave and put a gigantic rock in front of the entrance to make sure she didn't escape.

After that, men's and women's lives changed completely: fights, wars, no crops, people starving to death.

A poor farmer figured out that all these troubles would come to an end if they could free the Goddess of Peace. He tried to move the boulder but it was too heavy for one man, so he went in search of his neighbors to get some help. But the god Hermes, who was the Gods' ambassador on Earth, tried to talk him out of it by telling him that a man shouldn't get in the middle of fights between powerful gods; and he urged him to stop, don't offend them, don't get them angry.

But wars and miseries were growing every day, so the desperate peasant went in search of his neighbors to convince them to help. Every friend then convinced his friends, and so on, until all the poor farmers were solidly behind him, and I mean this was a lot of people.

One day in defiance of the Goddess of War they all gathered at the boulder that sealed the cave and started to push it, until they finally moved it away from the opening, and Peace was freed. With the return of the Goddess of Peace on Earth, like magic the fields started to flower again, all the wars ended, mankind could feed itself, and live in peace.

Every once in a while I turned around to look at Caroline sitting near her friends, and I saw that she was following the play. That's when I realized that she was also remembering Doc's lectures on the urgent need for mankind to sign a "peace treaty" with Nature. Or else, he said, "humans kaputt."

9.2 - Linked Systems

System Earth, as we've seen, is a very complex physical system. Its subsystems, Water, Atmosphere, and Soil, are constantly interacting with each other in complicated ways. In these interactions each subsystem exchanges energy with the other two, who then return it in different forms over periods of time ranging from a day or two to many years.

We say that Water, Atmosphere, and Soil are "linked" to each other: any change in the physical properties of one will influence the physical state of the others, exactly like two linked pendulums , where the energy of one, after a time, is transferred to the other, and vice versa. The biosphere is the delicate interface which connects water, atmosphere, and land not only through physical processes but also through biological ones. [Fig. 9.1]

The Sun - immutable, tranquil, far away - provides the energy for all these "games." The influence of human activity on these connections is still not fully understood. How does the presence of "Man" affect these connections?

Until about 150 years ago the presence of human beings had not significantly altered the connections among these subsystems. The advent of the industrial age has, however, drastically changed this "natural" scenario because great quantities of energy are needed to feed the machinery of production. As we've seen, in order for a machine to produce useful Work it must take energy from a heat source and must, inevitably, return a portion of it to a cold source.

Where can we find the energy to nourish the heat source? Certainly fossil fuels have been, and will continue to be, for decades to come, the simplest and most economical source of energy. But fossil fuels have a significant defect: in order to burn they absorb Oxygen and produce Carbon!

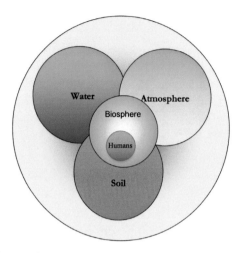

Figure 9.1 *The Earth-system.*

What is the cold source that receives the energy absorbed by the machine? Our environment! Since a machine's efficiency (the ratio between its mechanical output and the energy it needs to accomplish it) is rarely more than 30%, you can imagine how much thermal energy goes into the Biosphere!

9.3 Global Warming

By Global Warming we mean an increase over a period of time, a number of decades or so, in the temperature of the Atmosphere and the Oceans. This increase can have both natural and human causes. Such temperature change can be caused by either natural or human causes, Long term it will cause changes in many other phenomena which depend on temperature: average rainfall, humidity, cloud cover, atmospheric pressure, ocean salinity etc. In other words it will bring about Climate Change.[41]

Since the impact of Global Warming is worldwide, only an organization like the United Nations has been capable of promoting scientific, political, and economic research on the problem. In fact in 1988 two UN bodies, the WMO (World Meteorological Organization) and the UNEP (United Nations Environmental Program), founded a new organization called the IPCC (Intergovernmental Panel on Climate Change). The principal aim of the IPCC is to furnish the ruling bodies of the World with all the information necessary to understand the different causes and effects of climate change, including, in particular, the means to understand if and how human activity could be the cause of these changes. For this great and important work which it has realized (and continues to realize) the IPCC, together with Al Gore, has received the 2007 Nobel Peace Prize.

It is only in the recent past that we have reliable data on the average temperature of the Earth. From indirect, thus qualitative indications, we can extrapolate that in the last 1000 years – except for a short period of Medieval Warming (Greenland may have been colonized at this time), and a period of Minor Glaciation - the temperature of the Earth has remained relatively constant, with the exception of the last 100 years when we notice a noticeable increase. [See Fig. 9.2]

Basing their findings on a large quantity of data, particularly the more precise data collected in the last 100 years, the scientists of the IPCC in their Reports of 2001 and 2007 unfortunately confirm a warming of the globe.[42]

Figure 9.2 shows the data from the last 140 years. A constant increase in average temperature is noted from 1880 to 1940, a pause from 1940 to 1975, and a renewed increase beginning in 1975.

Using more recent evidence (see Reports from 2001, 2007, and 2013) the IPCC scientists have specifically concluded that:
  • the average surface global temperature (ie measured near the ground and ocean surfaces) has increased by

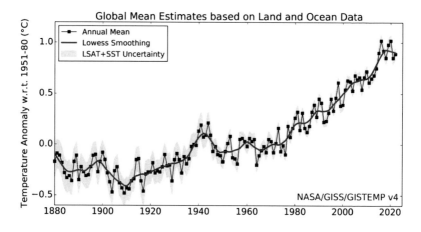

Figure 9.2 *Source, NASA-GSFC. The variation of the average temperature of the earth-ocean terrestrial surface since 1880 is current with reference to the average temperature between the years 1961 and 1990. The continuous black line is the annual global average and the red line is the average of five years . The gray shade represents the annual uncertainty total (LSAT and SST) in a 95% confidence interval.*

about 0.68 ± 0.2°C from 1880 to 2012; The largest increase (more than 0.5°C) has occurred between 1975 and 2010.

• the geographical distribution of temperature increase is not uniform, but has its highest value in the northern hemisphere. In addition, since 1950 the minimum night time temperature has increased more than the maximum daily daytime temperature.

• the period between 1983 and 2012 has in all likelihood been the warmest period in the last 1400 years.

What are the causes of this increase in temperature?

We've already stated that Earth's only source of energy is the Sun. The energy we receive from the Sun, whose average yearly value is about 342W/m², is not all absorbed by the Earth-system. A good portion of it, about

30%, is reflected back into space by the Atmosphere and the Earth's surface. The Earth's surface temperature depends on the amount of energy the Earth absorbs, in other words on the difference between the energy received and the energy reflected back. Since the energy we receive from the Sun, with minor variations due to solar spots, is pretty much constant, as is the distance from the Earth to the Sun, changes in the Earth's temperature depend solely on changes in the amount of energy absorbed or reflected.

An increase in the quantity of aerosols in the atmosphere also influences, to a lesser extent, how it reflects solar energy. Generally with more aerosols in the atmosphere more solar energy is reflected back into space, hence the temperature is lowered. But "black dust" produced by fossil fuels is an exception to this rule. This dust is a strong absorber of Infrared Radiation and therefore causes the temperature to rise.

The major cause of a decrease in reflected solar energy, and consequently an increase in absorbed energy, is an increase in the Greenhouse Effect in the atmosphere. This "beneficial" effect is controlled by the concentration in the atmosphere of the so called Greenhouse gases: Water Vapor, Carbon Dioxide (CO_2), Methane (CH_4), and Nitrogen Dioxide (NO_2). What has caused an increase in these aerosols? Humans! Our cities, our factories, our machines, our airplanes, our garbage, our fertilizers, etc.

By analyzing a large number of available data from the last two centuries, and from fossils and interior layers of glaciers, we have learned that since 1750, the beginning of the industrial revolution, CO_2 levels have risen by 31%, Methane by 151%, Nitrous Oxide (N_2O) by 17%. It's thought that these are the highest levels reached in 420,000 years. [See Fig. 9.3]

If it is true that global warming is due to an increase in greenhouse gases, which are caused primarily by human activity, then we should find that temperature increases in industrialized nations should be greater than in developing nations. And in fact the data demonstrates just that. [Fig. 9.4]

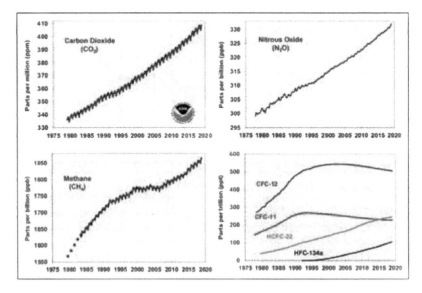

Figure 9.3 *Source, NOAA-ESRL. Atmospheric concentration of the principal greenhouse gases.*

The term Radiative Forcing is used to quantify the effect of Greenhouse Gases on climate. It represents the variations in the solar Energy Flux on Earth due to the increase in the concentration of these gases. Radiative Forcing is measured in W/m^2: a positive number indicates surface warming, a negative number surface cooling.

In addition to the increase in Greenhouse Gases concentrations, many other factors "force" Climate change. For example, Ozone concentration in the Troposphere and Stratosphere, alterations in the reflective properties of the soil due to changes in its use, the decrease in glaciers, and finally the slight variability in the solar energy flux.

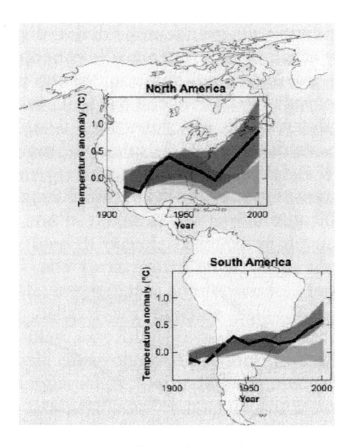

Figure 9.4 *Source, IPCC - Climate Change 2007. Comparison between observed changes in surface temperature (black line) in North America and South America. Natural causes are in blue, human and natural causes are in red. Note the general agreement between measured data and data generated by the human activity model.*

Figure 9.5 on the following page demonstrates the Radiative Forcing value relative to the concentration of each greenhouse gas. Because of insufficient data there is uncertainty regarding the contribution of aerosols.

Satellite data in particular (among other things) now provides us with clearer evidence of the consequences of temperature increase:

• From 1880 to 2015 sea levels have risen on average more than 20 cm. This increase is due to two factors: ocean warming and consequently an increase in water volume, and the melting of mountain and polar glaciers, particularly from Greenland and Antarctica.

• From 1971 to 2010 the temperature of the uppermost seventy meters of the oceans has increased at the rate of 0.11°C each decade.

• Mountain glacier surfaces in temperate zones have greatly decreased.

• The thickness and size of Arctic and Antarctic glaciers has greatly decreased. The permafrost in the northern

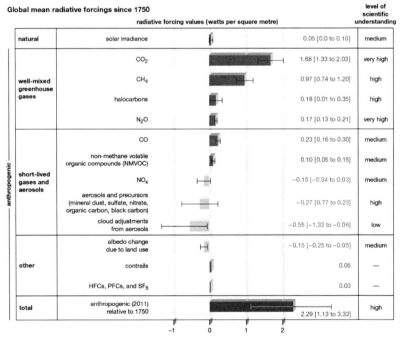

Global mean radiative forcings since 1750

		radiative forcing values (watts per square metre)	level of scientific understanding
natural	solar irradiance	0.05 [0.0 to 0.10]	medium
well-mixed greenhouse gases	CO$_2$	1.68 [1.33 to 2.03]	very high
	CH$_4$	0.97 [0.74 to 1.20]	high
	halocarbons	0.18 [0.01 to 0.35]	high
	N$_2$O	0.17 [0.13 to 0.21]	very high
short-lived gases and aerosols	CO	0.23 [0.16 to 0.30]	medium
	non-methane volatile organic compounds (NMVOC)	0.10 [0.05 to 0.15]	medium
	NO$_x$	−0.15 [−0.34 to 0.03]	medium
	aerosols and precursors (mineral dust, sulfate, nitrate, organic carbon, black carbon)	−0.27 [0.77 to 0.23]	high
	cloud adjustments from aerosols	−0.55 [−1.33 to −0.06]	low
other	albedo change due to land use	−0.15 [−0.25 to −0.05]	medium
	contrails	0.05	—
	HFCs, PFCs, and SF$_6$	0.03	—
total	anthropogenic (2011) relative to 1750	2.29 [1.13 to 3.33]	high

anthropogenic

−1   0   1   2

Figure 9.5 *Source: IPCC, climate change 2014. Average Radiative Forcing for the period 1750-2011. Estimates of the main contributions to the Global Mean Radiative Forcing for the period 1750-2011. The values reported are what in technical terms are called Effective Radiative Forcing values.*

hemisphere has begun to melt. As a result we can foresee an emission into the atmosphere of large quantities of CO_2 and CH_4.

• There has been an increase in severe meteorological events.

• There has been an increase in the El Niño events.

• Ocean acidity has increased, with severe consequences on sea life and biodiversity.

• Desertification, even in temperate zones, is constantly increasing. This is causing terrible social upheaval, and forced emigration of many populations.

• The increased migration of animals and plants from tropical zones to temperate ones is causing in a number of regions of our planet a significant weakening of biodiversity.

10

CLIMATE CHANGE

10.1 – Climate Change

Climate is the average status of meteorological conditions (temperature, air pressure, humidity, etc.) on a particular portion of the Earth's surface in a period of at least thirty years. For this reason any discussion of climate must necessarily use terms like *average* and *probable*.[43]

In order to help political decisionmakers with future strategies for their countries, scientists in the Intergovernmental Panel on Climate Change (IPCC) have gone well beyond verifying that the Earth's temperature is increasing, and are asking themselves whether it is possible (given our somewhat limited knowledge of certain global phenomena) to create models which can predict, with scientific methods, future climate on our planet and consequently our life on Earth in the ongoing century.

10.2 – Climate models

Climate models are constructed by applying the laws of physics to the different components of global climate. As previously noted, in the Earth system soil, air, water and the biosphere are linked together in such a way that any chemical-physical-biological change, even a small one, in any of these systems will eventually change the others.

In order to create a climate model for our planet we must first create models for the atmosphere, for the Earth's surface, for the oceans and for the biosphere. We must then evaluate all the possible interactions among these systems and finally verify if, and how, each model deviates from the data in our possession.

In order to predict the future of our climate we must then also create a model for anthropic (human) development since the industrial era, so that we can quantify and estimate the effect of human activity on the principal components of our climate. Taking into consideration geographic distribution, enormous disparities in wealth, sanitary and cultural conditions, such a model must consider our exploitation of natural resources, increasing pollution, land uses, agricultural methods, deforestation, urbanization, means of transportation, exploitation of natural and energy sources, and loss of biodiversity. Clearly the anthropic model is the most difficult to design since, unlike climate systems, which generally follow the laws of physics, human activity and development many times follows unpredictable laws.

Because of these difficulties, the IPCC scientists have hypothesized various scenarios of human development and greenhouse gas concentrations in the next hundred years. These scenarios are called Representative Concentration Pathways (RCPs).

The pathways describe different climate futures, all of which are considered possible depending on the amount of greenhouse gas concentrations over time. These different possibilities are used to calculate a range of radiative forcing values for our planet (as Earth absorbs energy from the sun it must eventually emit an equal amount of energy into the atmosphere. The difference between incoming and outgoing radiation is called radiative forcing).

A knowledge of the range of radiative forcing values will help us understand how in the coming years the temperature of the oceans and the air will vary compared to

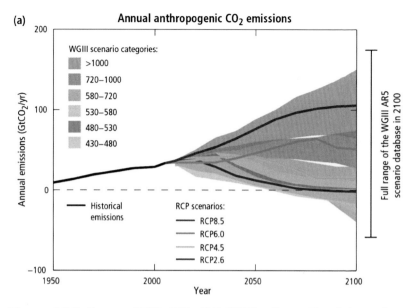

Figure 10.1 *Source, IPCC SYR_AR5_FINAL_ull_es_ Trend in carbon dioxide (CO_2) emissions in relation to the different RCP scenarios . Concentrations of anthropogenic CO_2 in the atmosphere in the past and projections for the future for the different scenarios.*

our pre-industrial past, and will consequently allow us to make projections about our climate future. Since, as previously stated, climate models are based on probability, with the passage of time and greater knowledge of these values, the models will become more and more accurate.
[Fig. 10.1]

The most pessimistic scenario, labeled RCP 8.5, foresees a constant and continuous use of fossil fuel energy which will cause ever increasing radiative forcing over time, reaching by the year 2100 more than 8.5 W/m^2 (watts per square meter), and then stabilizing at greater values by the year 2250. The most optimistic scenario, RCP 2.6, foresees an economy based mostly on service and information industries. It presupposes a 50% reduction in the use of fossil fuels from 2010 to 2030, and zero emissions by 2050. This scenario projects a maximum radiative forcing value of 3 W/m^2 by mid-century, and a continuing lowering to 2.6 W/m^2 by the year 2100.

Based on these two drastically different scenarios, we can make the following projections about our climate future:

• the continued emission of greenhouse gases, especially CO_2, will cause an increase in radiative forcing, the predominant cause of temperature increase, and climate change in the XXI century;

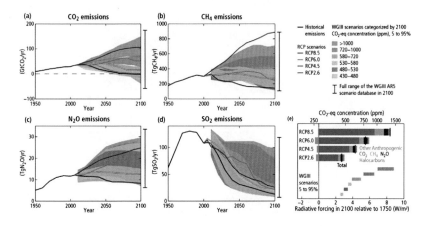

Figure 10.2 *Source, IPCC SYR_AR5_FINAL_ull_en_ Emission scenarios and resulting radiative forcing levels for representative concentration paths. Graphs a) to d) show emissions of carbon dioxide (CO_2), methane (CH_4), nitrous oxide (N_2O) and sulfur dioxide (SO_2). Graph e) shows future levels of radiative forcing.*

• in scenario RCP 2.6, CO_2 concentration in the atmosphere will increase from its current level of 440 ppm to 421 ppm by the year 2100;

• in scenario RCP 8.5, CO_2 concentration will increase from 440 ppm to 936 ppm by the year 2100; *Note that in the pre-industrial era CO_2 concentration was 250 ppm.

• all the RCP scenarios, with the exception of RCP 2.6,

predict that the average Earth's temperature will be more than 1.5°C higher than it was before the industrial revolution In the most pessimistic scenario (RCP 8.5) it may be greater than 4.8°C higher! [Fig.10.2]

The heating of our atmosphere will cause ocean temperatures to rise from their surface to their deepest

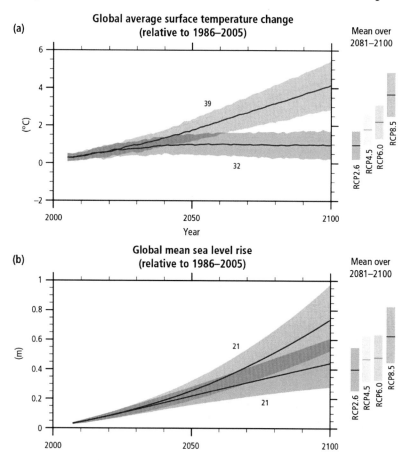

Figure 10.3 *Source, IPCC SYR_AR5_FINAL_ull_en_ Variation of Earth's surface temperature in the past and predictions for the future for two scenarios. Time series projections and uncertainty measurement (shading) are shown in relation to the RCP2.6 (blue) and RCP8.5 (red) scenarios.*

depths, with different and worrying results:

- ocean volume and ocean levels will rise;
- there will be a decrease in size and thickness of both ocean and land glaciers, including the melting of the permafrost in the northern hemisphere. In the most pessimistic hypothesis, it is possible that glaciers at the North Pole may disappear in the summer by the end of this century. The melting of land glaciers, particularly in Greenland and the South Pole, will increase sea levels even more;
- ocean evaporation will increase causing an increase in water vapor in the atmosphere, contributing to more and more "extreme" weather events;
- ocean temperatures may increase by anywhere from 0.6° to 2°C at a depth of one hundred meters which may decrease the speed of the oceanic "conveyor belt."

Sea level rise, caused by the combined effect of melting land glaciers and ocean temperature increase, is without doubt the most consequential event for humanity. As we can see in Fig. 10.3, depending on which climate scenario we use, the average global sea level increase from the period between 1986-2005 to the end of this century will be from twenty-six to fifty- five cm (RCP 2.6) or forty-five to ninety-eight cm (RCP 8.5).[44] The consequences of the latter increase will be dire for ocean bordering nations. Entire cities, regions, and even countries will be completely submerged. If we don't do something to defend them, many beautiful cities – think about Venice! – will disappear forever.

11

CONCLUSION

11.1 – Hopes

A fantastic myth about the creation of the universe asserts that Eros, the Greek god of Love, was hatched from a silver egg which had emerged from the void, and created the Earth.

It's not hard to imagine that the god of Love, whoever he may have been, "mixed" some water, carbon and calcium, and created man. It's also not hard to imagine that if Love doesn't force mankind to protect and conserve our planet's marvelous resources, the lives of future generations – also the fruits of Love – will be severely compromised.

11.2 – What to do?

For many years scientists have been warning us with incontrovertible facts that global warming, diminishing bio-diversity, and many other planet emergencies are caused by human activity. Each year is warmer than the previous one, each year more plastic covers the oceans, each year there are more poor people, each year there are fewer forests, each year there is less water to drink, each year...

Science has not only furnished us with the proof and causes of these disasters, it has also theorized, as we've seen, the impact they will have on future life.

We must first of all admit that we are living in extremely critical moments for our planet, and in order to

180

overcome them we must, as soon as possible, reduce to zero any increase in greenhouse gas concentrations in the atmosphere, so that at the very least current levels can be maintained. What can be done? There are so many things, but some are absolute priorities:

- there needs to be zero increase in the consumption of energy from fossil fuels, zero increase in the use of other sources of greenhouse gases (ex. appliances with old technologies, agricultural products shipped from distant lands, public transportation powered by fossil fuels, plastic containers, especially single use containers, etc.), by using less polluting alternative sources like wind energy, ocean currents, river flow, photovoltaic and geothermal energy;
- we must replenish our great forests and combat desertification. It is estimated that reforestation of the earth's surface would reduce CO_2 concentration by about 40-70 ppm (this effect is obviously not immediate since it takes about ten years for a new tree to become an efficient absorber of CO_2);
- we must improve agricultural and livestock rearing practices and methods in order to reduce greenhouse gas emissions and water usage;
- we must reduce "waste" and "wastefulness" (we use only 50% of what we buy, the rest becomes garbage);
- we must find ways to store emitted greenhouse gases.

These are only some of the principal things we must do, and they are the responsibility of both individual people and national and international institutions to do them. Taking these actions will not be easy or fast, since they require finding different models of social, economic and technological development from those currently in use. But we must implement them right now, for we are running out of time!

Numerous international governmental conferences, beginning with the one held at Kyoto, Japan in 1977, resulting in the Kyoto Protocol, have recognized that we must safeguard the wellbeing of future generations, but economic and political considerations have meant that government commitments are voluntary. For example, it is only in 2021 that the United States (after a change in Administration) has rejoined the Paris Agreement on Climate Change adopted in 2015.

In addition to political and economic factors, there is another important consideration to take into account: climate change inertia. Any variation in climate requires many years to take effect. For example, if we stop increasing the emission of greenhouse gases, the temperature of the atmosphere would not decrease, it would only stabilize. It would take a few thousand years for its temperature to decrease naturally. If we want to decrease global temperature more quickly we need to artificially remove greenhouse gases from the atmosphere.

During the Paris Conference (COP 21) in 2015 scientists from the IPCC outlined the urgent actions that must be taken in order to reduce greenhouse gas concentrations and limit temperature rise this century to below 2°C above pre-industrial levels, and in fac to limit increase even further to 1.5°C.

Limiting temperature increase to 1.5°C rather than 2°C would have important benefits for our planet: sea level rise would, on average, be ten centimeters lower than the rise at degrees; the North Pole might remain frozen even in summer; ocean water would be less acidic, limiting risk to biodiversity in the sea, and in particular saving many coral reefs; risks to biodiversity on land would be reduced; great danger to humans and animals, and indeed deaths, from "heat islands" would also be reduced.

11.3 – Mitigation and adaptation

As we've seen, the only possible way out of this grave emergency is to eliminate the consumption of energy derived from fossil fuels and the other sources of greenhouse gases, and to utilize alternative energy sources. Since obviously total elimination cannot be accomplished immediately, we must begin by minimizing the use of fossil fuels without interruption, so that, for example we might reduce greenhouse gas emissions to 50% by 2030 and 0% by 2050. These targets are compatible with the desired goal of keeping temperature rise to only 1.5°C by the end of this century.

The implementation of this philosophy of "mitigation" is not easy, since energy from fossil fuels continues to be the least costly, so that developing nations (and not only those) will find it hard to stop using it without "compensation" from wealthier nations. Industrialized nations meanwhile, though better able to afford alternative energy,

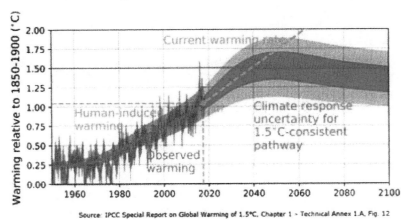

Source: IPCC Special Report on Global Warming of 1.5°C, Chapter 1 - Technical Annex 1.A, Fig. 12

Figure 11.1 *Source: IPCC, Special Report on Global Warming of 1.5°C. How close are we to a 1.5°C temperature rise? Answer: very close, by the year 2040! The yellow dotted line indicates how the temperature of the Earth's surface will increase if no serious policy to reduce CO_2 emissions is implemented.*

are also finding resistance to their use, especially by individuals and institutions who are ill informed or egotistically suffer from the culture of "not in my backyard."

The belief that we can implement policies of mitigation whenever we choose is a grave mistake. As we've seen, we can't simply immediately turn back the clock. It will take hundreds, perhaps thousands, of years to return to "adequate" greenhouse gas concentrations; so the longer we wait the longer it will take to get there. Because the Earth responds very slowly to positive as well as negative changes to radiative forcing, future generations will have no alternatives except to adapt.

Mitigation and adaptation are the key terms scientists and politicians in international conferences use when they describe actions necessary to be taken in the immediate future. What these two terms mean is, however, the subject of much debate among them. But it is clear that both actions must begin right now. Without immediate and intense efforts at mitigation future generations will simply not be able to adapt. The problem has become a "battle" between the present and the future; between us "adults" from the last century, and the "young" of the present and future centuries; between the current high costs of mitigation and future costs of adaptation - indeed of survival. It's currently an uneven battle, since the levers of power are still in the hands of the adults; but the young have the imagination, the will and the strength to find the best strategies to convince us that what we have been doing (thinking it was for our wellbeing) has been all wrong. The adults have brought us to the precipice. They must be reminded over and over again that their children and grandchildren do not want to fall into it. Will we be able to set aside our selfishness and sacrifice some of our privileges for the benefit of our descendants? Will we be able to make peace with nature before we perish?

A wise man once said that we do not inherit the Earth, we borrow it from our children. We must do everything possible to give it back in good condition, just as our children must do for theirs! [45]

NOTES

1.   In 2003 the Columbia shuttle burned up as it reentered the atmosphere due to intense heat caused by damage to a number of the ceramic tiles coating its shell. Tragically the entire crew was killed.

2.   See the web site http://orbits.eoportal.org/ to learn where to look in the sky.

3.   The Plank constant is 6.626 x 10^{-34} Js, where J (Joule)is the unit of measure of energy and s is the seconds.

4   Absolute temperature T is measured in Kelvin degrees; the relationship between T in Kelvin and the temperature in Centigrade t is: T = t+273.17.

5.   A body which is a good absorber of electromagnetic energy is also a good emitter, therefore a black body is the best emitter.

6.   1μm equals 1 millionth of a meter.

7.   A pre-Socratic philosopher born in Agrigento (Sicily) circa 492 BC.

8.   Recently the number of planets has been reduced from 9 to 8.

9.   Planets revolve around the Sun in elliptical orbits whose minor axis is called perihelion and major axis aphelion. The Earth's orbit is very similar to a circle because its two axes are almost equal; the perihelion is about 147,000,000 Km (91,000,000 mi) and the aphelion is about 152,600,000 Km (94,800,000 mi).

10.   The distribution of a black body.

11.   The energy hitting a surface of 1 square meter of a planet in 1 second is inversely proportional to the square of its distance from the Sun.

12.   The relationship between the energy reflected and that hitting a surface is also called Albedo.

13.   The great mathematician Frances J.B. Fourier was the first person, in 1827, to comprehend that the atmosphere behaves like a giant greenhouse trapping heat.

14.   Note that plants unlike animals do not need to ingest anything in order to feed themselves.

15.   See chapter 2.

16.   Diffusion also happens at other wavelengths but its effect is much smaller because it is inversely proportional to the fourth power of the wavelength.

17.   H.W.M. Olbers (1758-1840), German astronomer.

18.   The discovery of how many molecules there are in 30 grams of air was made about 200 years ago by a great Italian scientist: Amedeo Avogadro.

19.   There are numerous units of measure for atmospheric pressure:
- millimeters of mercury: mmHg (the atmospheric pressure at sea level under normal conditions is 736 mmHg, meaning that the weight of a parallelepiped filled with Mercury with a 1 square meter base and a height of 736 mm is the same as the weight of a prism with the same base and height of 600 Km which is filled with air);
- Hectopascal, hPa = 100Pa;
- Millibar, mb = 1hPa.

20.   At times it is assumed that water vapor consists of tiny droplets. This is not true, water vapor is a gas.

21.   UV radiation has a wavelength ranging from about 0.015μm (very high UV energy) to 0.400μm (low UV energy).

22.   When an atom of Oxygen encounters a molecule of Ozone it causes the reaction: $O + O_3 = O_2 + O_2$.

23.   These compounds are produced by the Oceans or the Earth's surface.

24.   In addition to CFCs , Methane (CH_4), Nitrogen Oxide (N_2O), Water Vapor (H_2O) etc., which are injected into the atmosphere by strong ascending currents, also contribute to the destruction of Ozone. These molecules are broken apart by UV rays in the atmosphere and become lethal destroyers of Ozone.

25   These events occur during the Christmas season or the birth of Jesus who is called "El Niño Jesus" in Spanish.

26. 1.5×10^{21} kg.

27.   This property is tied to the specific heat of water (the amount of energy needed to raise the temperature of one gram of water by 1°C - called a calorie) which is very high.

28.   Called thermal inertia.

29.   From a thermal standpoint there's a big difference between air and water. While the former can retain heat for a day or so, the latter can do so for years perhaps centuries.

30.   www.unesco.org/water/water_celebrations/decades/index.shtml

31.   www.greencrossitalia.it/ita/acqua/petizione.htm

32.   If a piece of ice floating in a completely full glass of water melts, will the water spill over? The answer is no (Archimedes' principle).

33.   The four fundamental forces of nature (known until now) are: gravitational force, electromagnetic force, weak nuclear force, strong nuclear force.

34.   NDVI is the new vegetation indicator which takes into account the effects of atmospheric aerosols and the effect of the ground cover under the leaf canopy. If we indicate with NIR the percentage of solar radiation reflected in the Near Infrared range (0.73-1.00 microns), and with VIS the percentage of solar radiation reflected in the Visible range (0.55-0.70 microns), then NDVI will be close to 1 when the surface is completely covered by healthy vegetation; close to -1 if the surface is the sea surface; and will approach 0 when the area observed is either desert or tundra.

35.   One liter is the volume of a cube whose sides are 10 cm x 10 cm x 10 cm. It equals 1,000 cm³.

36.   See Chapter 5.

37.   1 GtC = 1 billion tons of Carbon

38.   Rapa Nui was discovered by the Dutch explorer Jacob Roggenberg on Easter Sunday 1772. Hence the name Easter Island.

39.   The ruins attest to periods of great wealth, and a population of about thirty thousand at the height of its civilization.

40.   www.icun.org

41.   To speak of Climate Change presupposes that the average values of the parameters that define Climate have changed significantly in a few dozen years.

42.   Intergovernmental Panel on Climate Change, 2001: WG1 - The Scientific Basis. www.ipcc.ch/pub.

43.   "Probability" terminology has unfortunately created a mistakenly sceptical attitude toward climate change and global warming among many people, and in particular among politicians.

44.   There is confirmed evidence that ocean levels rose about 1.2 mm per year between 1880 and 1960. From 1960 to the present the increase has been about 2.5 mm per year.

45.   It has been reported that the secret services of the most powerful nations have concluded that the biggest threat in the future is not war but climate change!

About the Author and Translator

Vittorio De Cosmo was born in 1946 in Sepino, a beautiful hilltop town in the Molise region of Italy. He received a degree in Physics from the University of Florence in 1972 and has had a long-distinguished career as a university professor, researcher, and lecturer in various countries especially in the field of astrophysics. Before his recent retirement he worked for many years at the Italian Space Agency (ASI) where, among other duties, he was Head of the Earth Observation Section, and the Italian delegate to a number of commissions of the European Space Agency. He has authored numerous scientific papers and has participated in and chaired a number of national and international conferences in his field. This book is part of his continuing effort to instill in young people - especially middle and high school students - respect for the environment, and in particular a better understanding of the dangers our planet faces from climate change and global warming. His other current passions are his volunteer work for Lions Clubs International, trekking, music, and writing "climatic" fables for children. For the last thirty-five years he has lived in Rome with his wife Diana.

Marino D'Orazio (a close lifelong friend of the author) was also born in Sepino, Italy and emigrated with his family to the US at the age of eleven. He holds a Ph.D. in Comparative Literature from the CUNY Graduate Center; has been a college professor of both English and Italian, and has translated fiction and nonfiction books from the Italian. He has also had a long career as an attorney, from which he now deems himself 'semi-retired'. He has three grown children, six grandchildren, and lives with his wife in Saratoga Springs, NY.

Made in United States
North Haven, CT
29 April 2024

51901246R00108